Impossible Extinction

Impossible Extinction is about a remarkable journey. Every 225 million years the Earth, and all the life on it, completes one revolution around our Milky Way Galaxy. During this journey, life is influenced by calamitous changes. Comets and asteroids strike the surface of the Earth, stars explode near by, enormous volcanoes erupt, and more recently, humans litter the planet with waste. Many animals and plants go extinct during the voyage, but humble microbes, simple creatures made of a single cell, survive this journey. This book takes a tour of the microbial world, from the coldest and deepest places on Earth to the hottest and highest, and witnesses some of the most catastrophic events that life can face. This remarkable story for the general reader explains how microbes have survived on Earth for over three billion years.

CHARLES COCKELL is a microbiologist with the British Antarctic Survey, Cambridge, and the Search for Extraterrestrial Intelligence Institute (SETI) at the NASA Ames Research Center in Mountain View, California. His scientific interests involve exobiology, life in the extremes, and the human exploration of Mars. He received his doctorate from the University of Oxford and was Science Correspondent for the *Oxford Times* for three years, writing a regular science feature column. He has been on expeditions to the Arctic, Antarctic and Mongolia and in 1993 he flew a modified insect-collecting ultra-light aircraft over the Indonesian rainforests. He is Chair of the Earth and Space Foundation, a charity that supports expeditions that forge links between space exploration and environmentalism. He is currently President of the Mars Club, a dinner club for explorers of the martian poles, mountains, and deserts.

Impossible Extinction

Natural catastrophes and the supremacy of the microbial world

CHARLES S. COCKELL

CAMBRIDGE
UNIVERSITY PRESS

PUBLISHED BY THE PRESS SYNDICATE OF THE UNIVERSITY OF CAMBRIDGE
The Pitt Building, Trumpington Street, Cambridge, United Kingdom

CAMBRIDGE UNIVERSITY PRESS
The Edinburgh Building, Cambridge CB2 2RU, UK
40 West 20th Street, New York, NY 10011-4211, USA
477 Williamstown Road, Port Melbourne, VIC 3207, Australia
Ruiz de Alarcón 13, 28014 Madrid, Spain
Dock House, The Waterfront, Cape Town 8001, South Africa

http://www.cambridge.org

© Cambridge University Press 2003

First published 2003

Printed in the United Kingdom at the University Press, Cambridge

Typeface Trump Mediaeval 9.5/15 pt   *System* LATEX 2_ε [TB]

A catalogue record for this book is available from the British Library

Library of Congress Cataloguing in Publication data

Cockell, Charles.
Impossible extinction: natural catastrophes and the supremacy of the microbial
world / Charles S. Cockell.
     p.   cm.
Includes bibliographical references and index.
ISBN 0 521 81736 6
1. Microbial ecology.   2. Extreme environments – Microbiology.
3. Extinction (Biology)   4. Catastrophes (Geology)   5. Exobiology.   I. Title.
QR100 .C63   2003
579′.17–dc21   2002074046

ISBN 0 521 81736 6 hardback

Contents

Preface

Imagine an enormous ark, full of animals. On this ark is enough food for the inhabitants and the ark is on a journey. It's on a long journey that will take 225 million years before the animals arrive back where they started. If humans stayed on the ark for that long, they would pass through six million generations. But this is not an imaginary ark; it is the Earth on its journey once around the spiral arms of our Milky Way Galaxy, dragged unwillingly along by our master, the Sun. The animals on the ark are mostly quite unaware, but completely at the mercy of, the adventures that will unfold during the voyage.

During the 4.5 billion year history of the Earth, this voyage has been completed twenty-two times, but this book is about one of those journeys: a typical journey around the Galaxy.

It is very difficult for us to comprehend the length of time it takes to go once round the Galaxy. The life span of a human is short in geological terms. A person who lives 72 years will have the experience of traveling less than a millionth of the way around the Galaxy. If you think of the journey as analogous to a two-week holiday driving around Texas, then a typical person turns up for just less than half a second of the holiday. You can understand that they would have a very inaccurate view of the holiday and all the mishaps and adventures that might have happened during the full two weeks. The human species itself has only had the collective experience of passing through a hundredth of this galactic journey – three hours of our summer holiday.

And so our view of this 225 million year journey can be misguided by our short experience of life and our short collective experience as a species. However, over the last century we have begun to learn more about it. Scientists have turned successively more powerful telescopes to our galactic neighborhood. Through these telescopes they have seen the remnants of exploding stars and they have seen the deathly dance of two stars sucking the life out of each other and then exploding in a display of galactic fireworks, testament to the astrophysical violence that is commonplace in our Galaxy. They have studied the ancient and vast volcanic flood plains that litter the surface of our planet and begun to understand how the Earth itself, through the unpredictable eruptions of its restless molten core, can intermittently bring a rain of fire to life on Earth. The craters of asteroids and comets that have pummelled our planet in its past have been discovered and studied. Some of these are small, but some are so large that the energy released during the impacts is believed to have contributed to the end of some of the most spectacular creatures on Earth, including the dinosaurs.

Through these discoveries, we have, if you like, been presented with a slide show of the bits of the journey we missed. And the show has been quite a surprise.

In the *Rime of the Ancient Mariner*, Samuel Taylor Coleridge (1772–1834) tells of a journey; a ship beset by periods of calm interrupted by storm in a gruelling saga of seafaring, its crew observing, but for the most part powerless, to influence the course on which they are heading,

> *Day after day, day after day,*
> *We stuck, nor breath nor motion;*
> *As idle as a painted ship*
> *Upon a painted ocean*

But in an instant;

> *And now the storm-blast came, and he*
> *Was tyrannous and strong:*
> *He struck with his o'taking wings,*
> *And chased us south along.*

With sloping masts and dipping prow,
As who pursued with yell and blow
Still treads the shadow of his foe,
And forward bends his head,
The ship drove fast, loud roared the blast,
And southward aye we fled.

This book is about a galactic journey of calm interspersed by episodes of sudden catastrophe. It is about the adaptations one needs to have and the habitats one needs to live in to survive this journey. And in many ways it's a book that is dedicated to microbes, tiny one celled creatures up to fifty times smaller than the width of a human hair. As we will discover, these are the survivors of our journey, the ones who truly merit the biblical prophecy, 'The meek shall inherit the Earth'.

We'll discover that this is a remarkable story, one that is more than a saga of destruction, but of the quite wonderful resilience of life, the extraordinary interplay between biological evolution and the cosmic environment. It is, if you like, the story of Darwinism in the galactic context, the survival of the luckiest or the fittest organisms against cosmic selection pressures. These are selection pressures that derive their origins from the nature of the formation of the galaxies, stars and planets. Only recently has a species emerged that has the intelligence to recognize and study these perils and act on them. That is ourselves. But we are a new phenomenon.

As the story unfolds we will discover that most organisms are unfit to survive this voyage; they are transitory phenomena that, if they are not made extinct by their genes, are likely to eventually succumb to the perils encountered during the journey. Microbes have survived these dangers and in some cases they might have taken advantage of them to propagate themselves beyond the Earth. They are the survivors of a 4.5 billion year voyage and quite possibly, an impossible extinction.

1   The galactic roulette

Our lives are full of journeys, short ones and long ones, to our neighbors just down the street and our friends and cousins in other countries. But rarely do any of us consider the most important voyage that we taking part in; a voyage that takes 225 million years. And yet that is the time it takes our Solar System, the Sun and planets, to go once round our Milky Way Galaxy (Plate I). Even the Earth's orbit around the Sun, one year, is trivial compared to this immensely long galactic journey.

I recently came across a rather quaint email discussion between some amateur astronomers about this journey. One suggested that because we have a name for the time it takes for the Earth to go around the Sun (a "year") maybe we should have a name for the time it takes for the Solar System to go around the Galaxy. The email replies came in thick and fast. A "Gal-year" was one suggestion and a rather nice sounding "Milk-year" was another. You make your own decision.

This voyage began about four and a half billion* years ago when the Sun and the planets in our Solar System were formed. When life began on the Earth about half a billion years later it became an unwilling back-seat passenger in the journey. It became vulnerable to the explosions of dying stars, impacts of icy comets and rocky asteroids and the volcanic unpredictabilities of Earth's molten core during this galactic merry-go-round. To understand how we got into this journey and what might happen during its course, we need to know something about where we all came from.

---

* Throughout this book I will use the US rather than UK billion, that is, 1,000,000,000.

Some parents will tell their young children that babies come from under gooseberry bushes, but this isn't entirely accurate. Babies actually come from supernovas; explosions of stars that died during the formation of our Galaxy ten billion years ago. All of us are made from the atoms generated in these violent stellar explosions. Your parents merely assembled these atoms into you, but that was a much less energy-intensive task and so they shouldn't take too much credit for your existence.

The supernova explosions were part of the fireworks that heralded the formation of the Milky Way in a cloud of hydrogen gas. As with all good arguments about the Universe you are probably asking: but where did the gas cloud come from? Apparently from the Big Bang, a fiery explosion fifteen billion years ago from which all matter in the Universe emanated, including the cloud from which our Galaxy came. The origin of the Big Bang is one of those mysteries that is still being debated. My focus in this book isn't to start arguing about the merits of various cosmological theories about how the Universe began, as it has little bearing on what I want to tell you about and would take us on a great philosophical diversion. So if you will forgive my early humiliating surrender to this great question, let us start with the gas cloud, the cloud from which the Milky Way was born.

The cloud is like a billow of smoke that comes off a garden bonfire. All of us, on a fine summer evening, have sat and watched smoke coming off a fire or even a barbecue. You'll know that it is not completely regular. It has swirls or eddies in it. Smoke circles form and dissipate in many patterns, puffing out and curling in on themselves. Our galactic cloud is in some ways like this. The cloud isn't uniform and patches of smaller cloud begin to form inside the bigger cloud. The eddies begin to collapse into themselves because of gravity. As we watch the gas cloud we begin to see small blobs forming within it. These are the protogalaxies that will eventually turn into individual galaxies. One of those regions of collapsing gas is the Milky Way protogalaxy.

As the Galaxy cooled and coalesced so it began to rotate, like a spinning top. Pull the string on a spinning top and it tends to stay in one place, but the energy you put into it from pulling the string is now in the form of rotation as it whizzes round. So the energy originally imparted to our galactic cloud is partly contained within its spin. This is how we ended up on this journey around the edge of the Galaxy, the journey that this book is about. We are on the edge of a giant spinning top.

We orbit the Galaxy around a giant black hole – a star so dense that it sucks in light itself, giving it a black appearance. Although you can't see it through a telescope, the center of our Galaxy is out there in the night sky towards the constellation Sagittarius (Plate II). A rather sobering experience one evening is to take an astronomy book and locate the constellation Sagittarius and contemplate that beyond those stars is the center of the Galaxy, the point about which we are revolving on our long galactic journey. The next time humans will stand and look towards the middle of the Milky Way from roughly the same place will be in 225 million years time.

Like anyone who is curious about the neighborhood in which they live, you might be wondering where we are now. The Solar System's path around the Galaxy is not perfectly circular. If you imagine yourself looking down on the Galaxy from above you can see its Catherine wheel-like appearance (Figure 1), our distance from the center varies between about 27,000 and 31,000 light years as we go around. Right now we are moving inwards toward "perigalacticon", the point in the orbit closest to the center. As well as going round and round, the Solar System also moves up and down. If you imagine that you are now looking sideways on to the Galaxy at a spiral arm then we would be just above the middle of it, about 75 light years above the middle. We go up and down every 60 million years. We passed through the middle of the galactic arm about 2–3 million years ago and at the moment we are on our way towards the edge of our arm of the Galaxy.

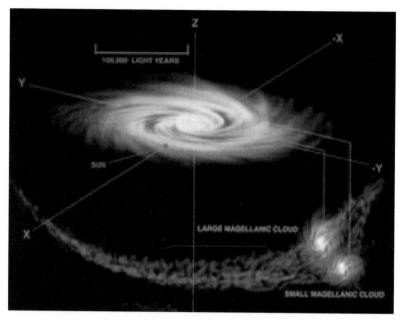

FIGURE 1. An artist's impression of our Milky Way Galaxy as seen from about 300,000 light years away. Also seen are the Magellanic Clouds, small irregular galaxies, in orbit around the Milky way. The position of our Sun is marked, and the X, Y, Z indicate the three-dimensional nature of the representation.

The Milky Way Galaxy is about 60,000 light years in diameter, about half a billion billion kilometers. To put it another way, if you started out at the edge of the Galaxy and drove to the other side at 50 miles an hour, your journey would take about a trillion years, much longer than the supposed lifetime of the Universe, so these are enormous distances. We are on the edge of an extraordinarily large spinning top.

We don't truly know what the galaxies looked like at the beginning of the Universe. Like the ancients and their view of the Earth as a disk on the back of a giant turtle, the ways we picture them are likely to be quite incorrect, a product not of our stupidity, but of the finite knowledge we have of our surroundings.

MORE ABOUT OUR LOCAL NEIGHBORHOOD

Our propensity to name places where we've been doesn't stop at holiday destinations. We are in a spiral arm of the Galaxy that we've named the Orion arm. The spacing between the spiral arms is about 4000 light years. Our neighborhood isn't exactly a throbbing hub of galactic social life, it has to be said. The density of stars is 1 star in every 300 cubic light years, with the nearest one, Proxima Centauri, 4.3 light years away or 41 trillion kilometers away, so that's quite a lonely existence. The history of our local neighborhood is actually quite exciting, so we've got that going for us. The Solar System is on the edge of an expanding shell of hot plasma about 400 light years in radius, which appears to have originated from multiple supernova explosions in a region called the Scorpius–Centaurus OB association. The shells of hot gas blown off the stars in the supernova explosions sweep material in front of them. Within this "Local Bubble", our Solar System is within a small interstellar cloud, 12 light years across, known as the Local Interstellar Cloud.

When the Hubble Space Telescope was launched into Earth Orbit in 1990 it began to yield remarkable images and information about our Universe. The telescope, being a machine, can be left to gather light from one spot in the sky for days at a time. During these long exposures it can gather pinpricks of light from galaxies that would

MORE ABOUT LIGHT YEARS

A light year is the distance traveled by light in a year. It is a confusing measurement because to the untrained eye the word "year" suggests a time period not a distance (some bad movies talk about things taking so many light years). A light year is actually 9.56 trillion kilometers. A light second is 300,000 kilometers. If you live in San Francisco and your aunt in Los Angeles, or if you live in London and your aunt lives in Edinburgh, you need to travel a two-billionth of a light year to visit her.

be impossible to see with the naked eye. The light from the galaxies seen in the Hubble Deep Field Views has taken so long to cross space and reach the Earth that we are now seeing the light given off by these galaxies during the early history of the Universe. In these images are the most primitive galaxies we have observed. Some of the furthest away, small red spots on an exposure, may have formed about fourteen and a half billion years ago. So here we are, at a stage in history where we can take photographs of the night sky from light that set out on its journey about eleven billion years before life even evolved on Earth. Imagine a photon of light formed in the gas cloud of a galaxy almost fifteen billion years ago, its ultimate destiny to hit the camera of a telescope orbiting a planet whose occupants seek to know their place in the Universe. This is the stage of human history in which we live.

Our Galaxy isn't alone in the void of space, an isolated cloud of stars. Because the protogalaxies were all forming within the gas cloud from separate swirls we would expect that there would be areas in the Universe with galactic clusters and other areas of the Universe that are quite devoid of them. This is actually what we observe. We don't find galaxies uniformly distributed, we find them collected into clusters, although the theories behind how exactly these clusters form are still quite controversial.

Humans are quite a sociable lot when it comes to our local neighborhood and we've gone to great lengths to find out who our neighbors actually are. We've called our own cluster of galaxies the "Local Group", made up of about twenty-seven galaxies and star formations of various shapes and sizes. "Local" is a relative word, of course, because this neighborhood is about three million light years across (that's thirty billion billion kilometers). The most spectacular galaxy, aside from our own, is Andromeda, another spiral galaxy about two million light years from us. Standing by our side are the Magellenic Clouds, small irregular shaped galaxies that appear to be in orbit around the Milky Way. In Figure 1 you can see an artist's impression of a view from deep in space looking back at our home.

MORE ABOUT TRAVELING LONG DISTANCES

The distances between stars and galaxies are so large that some scientists have invoked them as a reason to explain why we have not been visited by extraterrestrials. The question "If life is common in the Universe, why are we not visited by extraterrestrials?", called the Fermi paradox, is answered by the fact that distances in and between galaxies are so large that they are impossible to cross in reasonable time periods by any civilization. Hence we are bereft of friendly social visits from other civilizations. Others disagree with this and cite a number of other reasons why we don't see aliens, including that we are a scientific experiment to be observed, but not meddled with, or that we have already been visited, but can't recognize it.

On the edge of the Milky Way Galaxy is a small star that harbors our Solar System, the Earth and us. How did it come to be?

Processes in the Universe occur on the grand scale and on the tiniest scale. As the Galaxy began to form amongst the giant cluster of galaxies around it, so the galactic cloud itself began to coalesce into yet smaller fragments. These small collapsing gas clouds get hot in the middle and the radiation generated by the hot interior begins to force outwards until it resists the collapse caused by gravity pulling inwards. In the middle of this gas cloud a baby star begins to form.

Star formation happens all over the newly formed Galaxy. Everywhere where enough gas can coalesce and collapse to reach a high enough temperature for the nuclear fire to ignite, a star forms. In our Galaxy alone there are now 100 billion stars, all of which began from the swirling collapse of interstellar clouds. Our Sun is just one of these. You can only see about 2,500 stars with the naked eye on a clear night from any location on the surface of the Earth, but their white color and the bands they form is how "galaxy" got its name, from the Greek *"gala"*, meaning milk. About a quarter of the mass of the Galaxy is estimated to be in these stars, but galaxies are not very clean places. Although all we can see in the night sky are the pinpricks of light from bright stars, there are many other things out there as well. Just less than a quarter of the mass of the Galaxy is locked up in the remnants of old stars, another quarter is in interstellar clouds and then the rest is in "dark" matter. We don't know what dark matter is, but we need to hypothesize its existence to explain the stellar motions and the galactic journey that we are on. Speculation abounds about the nature of the dark matter, which includes everything from brown dwarfs, which are large objects that didn't ignite into stars, to undiscovered nuclear particles.

As the stars in the early Galaxy began to form, the temperature within them began to rise as they collapsed under gravity. The centers of the stars got so hot that protons, which are positively charged particles that came from hydrogen inside the galactic cloud, can collide with each other and form deuterium, a heavy form of hydrogen. Now when deuterium forms, it loses some mass. Albert Einstein's (1879–1955) famous equation, $E = mc^2$ explains why. You may think I'm going into too much detail, but all the equation says is that mass (m) and energy (E) can be interconverted (c is the speed of light). So when the deuterium breaks up and looses mass, this mass is converted to energy and released. What has essentially formed in our protostar is a nuclear fusion reactor and it starts to release great quantities of energy. The collision of deuterium with another proton forms helium and more mass is converted to yet more energy. Once the nuclear fire is ignited then we can truly say that a star is born.

Within this intriguing life story there is a mystery. I told you that the Galaxy and stars came from a hydrogen cloud (which maybe had a little helium mixed in as well), but you and I know that there is more in the Universe than hydrogen. There's calcium in our milk and iron in our fridge magnets. So where did these other elements come from, the elements from which life would emerge?

Imagine a hermit in an isolated log cabin in the countryside that is heated by an oil stove. One cold, harsh, winter the hermit runs out of oil and he doesn't have any money to buy more. The hermit might scour his house for something to burn to keep warm; maybe bits of wood from cutting up the furniture. Eventually as he got more desperate he would turn to successively different forms of energy to keep warm; perhaps burning bits of wet branches from trees and maybe even his old carpet. Finally, he would run out of things to burn and our poor hermit starts to go cold.

In a similar way, stars begin to run through different energy reserves. They start with hydrogen, and then move to helium. As this process goes on atoms fuse together to form heavier and heavier atoms to make the fuel that the star is burning. Interesting things are going on. Carbon might fuse with helium and form oxygen – the gas that you'll need to breathe. Oxygen atoms can fuse together to form sulfur and phosphorus – and there's your lawn fertilizer. And so, through this process, the various heavy elements we are familiar with are formed in the star. All the elements needed to make life are formed in this way. Everything that is in your house, every object you take for granted including yourself, owes its existence to the fusion reactions occurring inside the stars. The idea sounds quite simplistic, but remarkably the abundance of elements in the Universe seems to be quite well predicted by this idea. We find hydrogen is the most abundant element, followed by helium, carbon, oxygen and so on.

During the formation of these heavier and heavier elements, the core of the star begins to contract and the radiation pushing out can cause a large envelope to form around the star – a red giant is born. Inside the red giant is the core of the small star, steadily contracting as it runs through to heavier elements to burn. This little star is called a white dwarf. The mass of such a star can be formidable. If you could get hold of a tablespoon of a typical white dwarf, it would weigh about a metric tonne.

Eventually the star will use up its fuel resources and will slowly burn out like an old fire. Now and then, if the star was big enough in the beginning, the collapse of the star can be very dramatic. It can trigger a massive explosion. The explosion blows off the outer material from the star – a supernova has occurred. The star has to be just the right size to erupt into a supernova, bigger than one and a half times the mass of our own Sun. (Because this limit sets the size at which a star will go supernova, it turns out to be quite important for life because supernovas that occur near the Earth could have some dangerous consequences for life as we will see later in the book.) The stars and planets that have formed during the last five billion years, including our own Solar System, came from the material blown off by early supernovas in the Galaxy.

Today, supernovas are less common that they were during the formation of the Galaxy. It is supposed that there could have been one supernova every year in the original Milky Way and now there is less than one every ten years, possibly as few as one every one hundred years. During your lifetime there is a very good chance of at least one supernova exploding somewhere in the Milky Way, although it is unlikely you will see it. Sometimes they can occur close enough to be very spectacular. In the year 1054, the Chinese, Japanese and Koreans observed a star that was visible in daylight for three weeks and it was six times brighter than Venus in the night sky. This supernova, more formally cataloged SN (supernova) 1054, occurred in our Milky Way, and the remains are almost certainly the Crab Nebula, first observed in 1731 by astronomer John Bevis (1695–1771). It is a wonderfully colored nebula with a small rotating star in its center that sends out a pulsating signal by rotating 33 times a second. (Figure 2.)

By watching the rate at which the star is slowing down and then back-tracking, astrophysicists can work out roughly the date when the supernova explosion in the Crab Nebula was observed from the Earth. The date is about 1,000 years ago, which agrees with the observations of an explosion in 1054. (Remember though that this wasn't when the explosion itself actually occurred! Because the supernova is 6,500 light years away it took 6,500 years for the light to reach the Chinese observers. The explosion itself actually happened in about 5,500 BC, but was observed on the Earth in AD 1054.)

In 1953 at White Mesa and Navajo Canyon in Arizona, USA, an archeologist, William Miller, found Native American cave paintings dating to the beginning of the second millennium that show a bright

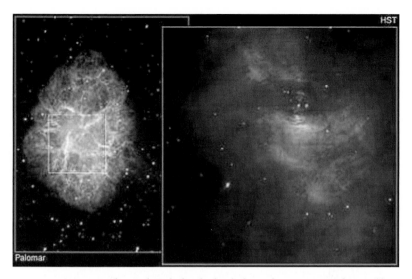

FIGURE 2. The Crab Nebula, the birthplace of a supernova observed by
the ancients in AD 1054. It is from these types of explosions that the
heavy elements necessary for life were formed and these explosions
themselves can threaten life. On the left is the Crab Nebula as seen
from Earth. On the right the Hubble Space Telescope peers into the
center of the nebula and gives us an image of ripples from the pulsar, the
remnant star formed during the explosion. (Image, Jeff Hester and Paul
Scowen, Arizona State University and NASA.)

star next to the Moon. The position of the Crab explosion, which
has been calculated to have occurred within 2 degrees of the moon,
fits the cave paintings. Luckily the supernova wasn't close enough to
have affected life on Earth. The radiation from it had diminished to
insignificant levels by the time it had traveled 6,500 light years. It was
a spectacular warning, unappreciated by the ancients, of the power of
these explosions occurring around us as we rotate around the Galaxy
on our journey.

I might have given you the impression that stars begin their
lives, burn through their fuel, and then, if they are big enough, explode
as supernovas. This is basically right, but the process can take a very
long time. As we've already seen, they start off their lives as bright,
fiery, tempestuous things, but once the nuclear furnace is ignited and

given a few million years, they settle down into adult life and become a little less erratic. This period of adult life, called the *main sequence,* takes up about three-quarters of a star's lifetime.

Because the main sequence takes up most of a star's lifetime before it burns out or turns into a supernova, a snapshot of our Galaxy will reveal that the majority of stars will be in this stage. Over 90% of the stars we observe in our Galaxy are in the main sequence. In some ways, you can think of this as like our society. Because the adult portion of our lives takes up more time than our childhood years, at any given time a snapshot of society will reveal that most people are adults. And for stars this adult life can go on for quite a while. Our own Sun is about 4.7 billion years old and will stay in the main sequence, probably for about another five billion years.

Oh Be a Fine Girl, Kiss Me Right Now or Soon. This splendid mnemonic, although probably one that is not very politically correct these days, is an easy way of remembering that not all stars are born equal. The different star types, OBFGKMRNS, a quite illogical series of letters, reflect a decreasing surface temperature of the star from left to right. The temperature of the star will depend upon its age and how much mass it had when it started. Our own Sun, which is a G-type star and has a temperature of about 6000 °C, is less common than the cooler K-type stars found in the Galaxy that have temperatures of about 4,500 °C. You'll remember that I told you that stars that turn into supernovas tend to be the larger stars that exceed the limit needed to make them violently collapse. These are often the O- and B-type stars. So stars not only have a life history, but like people, they also come in all sorts of sizes and vary in their fieriness or placidness. Astronomers Henry Russell (1877–1957) and Ejnar Hertzsprung (1873–1967) were the first to propose a graph of this life history, a graph that obviously then became known as the Hertzsprung–Russell diagram. It is in many ways the most basic and essential classification system for anyone interested in the evolution of stars. All stars follow the life story it represents, including our own.

Of the future of our own Sun in this emerging picture of stellar evolution we are not quite sure, but it is believed that in about five billion years it will leave the main sequence and begin to expand into a red giant. First the innermost planet Mercury will be vaporized, then Venus, already a hot cauldron of volcanism bubbling away at 464 °C. The oceans of the Earth will begin to boil and life on the land will be extinguished as the planet becomes fit only for heat-loving microbes. Eventually the Earth itself will boil and vaporize and at this point all life on it will be extinguished. Until this time the Sun and the Earth will continue on their galactic journey. We will complete another twenty-five rotations around the Milky Way.

2   **Primordial leftovers**

A star sixty-three light years away has given us a view of what our home was like when the Earth first cooled from the material orbiting around the early Sun. Around the star Beta Pictoris, see Figure 3, a disk of material has been seen. The star is surrounded by a dust disk so large that it extends to a distance equivalent to several times the distance of the Sun to our furthest planet, Pluto. The disk has revealed itself to be more than just a collection of dust and rocks. Near to the star there is a ripple-like appearance. The explanation for the ripples is the presence of one or more large planets, perhaps large gas planets. Beta Pictoris was the first really good example of a disk around a star that gives scientists a picture of how our own Solar System and the Earth might have looked those four and a half billion years ago when they first formed. The dust cloud, which flattens out like a disk around the star, doesn't contain all that much matter. In our own Solar System only about one percent of all the mass is contained within planets, the rest is in the Sun.

It's obvious from the dust cloud around Beta Pictoris that there is a lot of debris formed during this process. Like any good home improvement job, there are bits and pieces left over. In our Solar System, it is believed that a ring of icy rocks beyond the orbit of the planet Neptune encircles our Solar System. The rocks form the hypothetical Kuiper belt, named in honour of Gerard Peter Kuiper (1905–1973), a Dutch-born American astronomer who first suggested that this could be the origin of comets. Another Dutch-born astronomer, Jan Hendrik Oort (1900–1992), suggested that the Kuiper belt extended into a halo of icy rocks around the Solar System, a halo that is known as the Oort cloud.

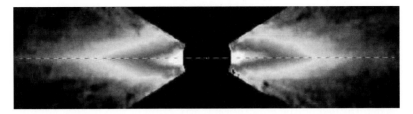

FIGURE 3. Around the star Beta Pictoris, at a distance of 63 light years, in the constellation Pictor, is a disk of dust seen here fanning out across the picture. The picture was taken with the Infrared Astronomical Satellite and the black area is just a region blocked out by the instrument. Within this disk might be one or more gas giant planets. This is how our own Solar System might have looked in its youth. (Image, NASA.)

Every now and then the passage of a star close to our Solar System as we undertake our journey around the Milky Way might disrupt this halo of debris and a few of the ice blocks and rocks might fall inwards, drawn in by the gravity of the Sun and the giant planet, Jupiter. Jupiter is so massive and has such a powerful gravitational influence that it will stop some of these objects before they penetrate the inner reaches of the Solar System, either sucking them up or ejecting them from the Solar System altogether. Some planetary scientists even speculate that without Jupiter, Earth might never have been able to sustain complex life because of the incessant bombardment of our planet by errant rocks from the outer Solar System. Having said that, planets that are regularly bombarded with these ice blocks and rocks might be more efficient at dispersing microbial life amongst the planets and stars too, so from a biological point of view we shouldn't become too anthropocentric in our ideas about the dangers of impacts.

Those pieces of icy rock that make it through to the inner Solar System get close enough to the Sun for the ice on their surface to begin to heat up and vaporize, giving rise to coma or vapor trails. These are the tails of comets that blaze rare streaks across our night sky. There is one comet that is particularly famous and rather nicely illustrates the predictive power of scientific enquiry. That comet is Comet Halley.

The great mathematician and originator of many of the fundamental laws of gravitation, Sir Isaac Newton (1642–1727), first suggested that if comets were made of matter they must be attracted to the Sun by the laws of gravity. He proposed that the path of a comet should therefore be an ellipse, parabola or some variation on a conical section as determined by the gravitational attraction of two bodies. Edmond Halley (1656–1742) reviewed the historical information on comets and concluded that the great trails of white observed in the skies of 1533, 1607 and 1682 were all the same comet. Using Newton's mathematics, he predicted its return in 1758. Sadly, Halley died in 1742 and so never lived to see this remarkable prediction come true, but in 1758 it did indeed return and hence the naming of Halley's comet.

The prediction Halley made is also one of beauty from the point of view of biology, because it demonstrates that the fate of life in the Universe is not an entirely random one. Predictable patterns of comet movements, the birth and death of stars and the movement of the Earth around the Galaxy might lead to periodicity in the effects of the cosmic environment on life. As we will see later in the book, the nature of this periodicity, particularly with respect to mass extinctions, is a hotly debated topic. Some see periodicity, some see pure random statistics at work with little evidence of a regular periodic effect of the cosmic environment on life. And so in some ways Halley is more than just a comet finder. He should really be credited with having provided the first piece of evidence for periodicity in objects capable of bringing destruction to life on Earth; the first person to provide evidence that over two centuries later would spark a fierce debate about the statistics of extinction.

Some of these left over pieces of ice and rock are nearer to us than the Oort cloud. The leftovers that cause the greatest excitement, not just amongst planetary scientists, but more recently amongst politicians who are worried about collisions, are rocks that cross the orbit of the Earth, often known as Near-Earth Objects. Mapping them and their orbits has become something of priority recently, as our eyes have opened to our possible vulnerability. A proportion of these comets and asteroids could eventually intersect with the orbit of the Earth and collide with the planet, causing massive impact events that disrupt life. In the past, some of them almost certainly have.

Although this book is about what happens to Earth as it journeys around our Galaxy, and we have already seen where supernovas and asteroid and comet impacts come from, it is important to understand why it begat life in the first place. The fact that our Earth supports life and sits between fiery Venus and frozen Mars doesn't seem to be coincidence. Our planet is sitting in what planetary scientists have called the *habitable zone,* an imaginary band around the Sun where liquid water is stable and life can exist. A planet that transgresses the inside of this band will suffer a runaway greenhouse effect, as Venus will attest. The water evaporates causing the planet to heat up. The gas you normally find in your fizzy drinks, carbon dioxide, is also released into the atmosphere from the rocks, where it is bound up as carbonate compounds. Because carbon dioxide is a greenhouse gas, it further contributes towards a warm up of the planet until the planet boils dry. The runaway greenhouse effect may not have been a feature of Venus for its entire life. Early in the history of the Solar System, when the Sun was less luminous than it is today, it is possible that Venus had hot oceans. The greenhouse effect may have taken hold several hundred million years after its formation, leading to the uninhabitable hell of a planet observed today.

The fate of life may also be sealed for planets that transgress the outer bounds of the habitable zone. Here the water will freeze and life will be denied the liquid water that is presumed to be essential for its existence. The carbon dioxide that might otherwise behave as a greenhouse gas and help warm the planet will also condense out of the atmosphere, further contributing to the freeze-out. Luckily our planet is within the thin band around the Sun which is at the right position for liquid water to exist.

There are some exceptions to this rule of the habitable zone. Europa, Jupiter's moon, is a case in point. Here tidal forces from the massive planet Jupiter contort and buckle the moon causing it to heat up and melt the water under an ice sheet. And so on a moon that is too far away from the Sun for solar energy to make liquid water a giant ocean is found. And there are other ways in which liquid water can be created on a planetary body without it having to be close to a star. Imagine a planet drifting through the blackness of space, on its own and not even bound by the gravity of a star; perhaps in its early history it was ejected from a Solar System by the close passage of another star. One can imagine that radioactive elements in the core of the planet might heat it up, providing just enough heat to melt ice and provide a habitat for life. So the habitable zone is in some ways a misleading concept, but it is useful because it helps us to understand how the cosmic environment influences our planet as an abode for life.

Even our Moon has been claimed as an important factor for the emergence of life on our planet. Without it, some say, the axis of the Earth would shift wildly over tens of millions of years, causing climatic changes that would make it difficult for complex life to evolve. By stabilizing the axis, the climate is also stabilized and provides a cradle for life to evolve. I have to admit to being skeptical of these types of arguments. Life has survived many quite large-scale climatic changes in the past and humans themselves have survived ice ages. It is difficult to say what the effects on evolution might be if the Earth did not have a moon. Below I have reproduced a paragraph from an imaginary book about evolution belonging to an intelligent species living on an Earth without a moon,

> Intelligence evolved on Earth because, unlike some other planets, we have no moon. If we had a large moon the axis of our planet would be too stabilised. The lack of a moon allowed relatively rapid (over millions of years) climatic changes to occur that provided a selection pressure for tool-building abilities and mental versatility that were required for complex animals to be able to rapidly adapt to these changes. As a result, intelligence evolved. On a planet with a large moon, life would not be forced to be so adaptable and it would suffer from greater evolutionarily stasis.

Hence, intelligence probably would not have evolved. This provides us with an example of how complex life and intelligence is made possible by the unusual characteristics of our planet.

One could even speculate that rapid changes in climate caused by changes in the tilt of the axis would create an evolutionary selection pressure for generalist organisms capable of surviving other sudden unexpected climatic perturbations such as those caused by asteroid and comet impacts or changes in sea level. By stabilizing the Earth's tilt so as to reduce climatic changes, the Moon, if you like, has made life too specialized to invariant habitats. The Moon has therefore made life more prone to mass extinction.

However, these types of speculations are scientifically problematic. Because the nature of terrestrial life has been determined by the physical conditions of the Earth, past and present, it is easy to become convinced that without those conditions life as we know it would be impossible. We don't know what alternative conditions could lead to complex life. The Earth is a sample size of one and any scientist will tell you that developing hypotheses with a sample of one is unwise.

Despite being in the habitable zone and having a stabilizing moon, other factors about our situation are not necessarily conducive to life. The rise of oxygen, which is essential for respiration (breathing) and thus, it is presumed, to complex animal life like us, apparently came about two billion years ago. It is produced by plant life and a variety of microbes that use sunlight as a source of energy – photosynthesis. Early in history, compounds like hydrogen, produced by volcanoes, constantly mopped up the oxygen produced by photosynthetic organisms. Only when these compounds were used up could oxygen begin to build up in the atmosphere, thus allowing complex life to evolve. It has been suggested that on planets where the source of these volcanic and subsurface gases is much smaller (perhaps because the planet is smaller), the rise of oxygen might be a lot more rapid. Thus, small, less-active planets might be able to shrug off two billion years of oxygen-free youth which constrained the evolution of life on the Earth and leap directly into an oxygen-rich environment fit for complex organisms perhaps meaning a more rapid rise towards intelligence.

Thus, to presume that the Earth is a planet where everything was "just right" for complex life, the one-in-a-billion place where a Goldilocks of factors came together, is not necessarily correct. It is true to say that with the knowledge we have of early Earth we can conceive of changing some characteristics of our planet that might even have accelerated the evolution of complex life and intelligence.

It isn't known how typical our Solar System is. Are all star systems revolving around the Galaxy something like ours? If there was life on other planets would it occupy the same sort of place in a Solar System and be frustrated by asteroid and comet impacts? Humanity is like a young child that really only has the experience of its local neighborhood. In recent years astronomers around the world have begun to expand their view by taking a look at planets around other stars. Remarkable efforts have been made to find planets around distant stars and the efforts have paid off handsomely. Over seventy extrasolar planets revolving around other stars were known at the beginning of the twenty-first century. Few of these planets are like our own. Hot Jupiters – giant planets like Jupiter revolving close to their parent star, have been found. A companion to the star 51 Pegasus is about half the size of Jupiter and revolves around the star in about four days, compared to Jupiter's fourteen years to go round the Sun.

Rather quickly a picture is emerging of a variety of quite alien star systems. Imagine living on a planet where every evening a giant planet like Jupiter rises up into the night sky, like a moon, with twisting swirls of reds, oranges and blues. Giant and ferocious atmospheric storms in the planet's atmosphere that could swallow a planet many times the size of your own are the views under which your civilization is born. How many of these star systems support planets with conditions like the Earth is a mystery at the moment, but the search for Earth-sized extrasolar planets has now become a focus of great scientific interest. If and when such planets are found we might be able

MORE ABOUT LOOKING FOR LIFE ON DISTANT PLANETS

The high concentration of oxygen in our atmosphere (about a fifth of the atmospheric composition) is primarily caused by photosynthesis, a light-driven process in some bacteria and algae, and all plants, that generates complex organic compounds and frees oxygen gas from water and carbon dioxide. It is a signature of widespread life on our planet because there are no known geological processes that can produce this quantity of the gas. In the absence of compounds like hydrogen from volcanoes, which would mop up the oxygen, the gas builds up to cause this gross planetary-scale imbalance. Because oxygen absorbs certain wavelengths of the electromagnetic spectrum it leaves a trace in the spectrum of light and heat given off by a planet. Astronomers could search for oxygen by examining the light given off from the atmospheres of distant planets orbiting other stars. This would allow them to diagnose life on a planet without directly seeing it. Some even suggest they could go further and look for artificial greenhouse gases, which cannot be produced in large quantities by natural processes. Compounds like CFCs in the atmospheres of extrasolar planets would suggest the presence of intelligent civilizations.

to deduce more about the conditions on early Earth and astronomers could look for the signatures of life on these planets. They could begin to examine what the galactic journey is like for other planets similar to our own.

The preponderance of giant planets circling other stars does, however, fit in with what is observed about the Galaxy. About half the stars in our Galaxy are part of binary star systems, where two stars revolve around each other. It seems that when the nebula collapses to form the first stars, a common outcome is that it breaks into two parts and two nuclear furnaces are independently lit up if there is enough mass to begin with. The large planets we can now observe around other stars may just be collapsed remnants of the cloud that didn't coalesce or get hot enough to fire the nuclear furnace. Astronomers calculate that if Jupiter itself were about one hundred times its present mass it, too, would ignite and turn into a star. We would now be living on a planet with two Suns, if, of course, we had come into existence at all in such a Solar System.

The purpose of the foregoing discussion is to fill you in on a little more detail about the Earth. We can understand how the Galaxy, stars and planets formed, but a little extra investigation is needed to understand how this planet in particular came to be able to support life. Some of the factors that allowed life, such as the habitable zone, do not bear directly on my discussion about extinctions, catastrophes and the microbial world, but they help us to understand life in the cosmic context. What nuances of galactic and stellar evolution have led to the life on our planet in the first place? What cosmic factors are important for our existence and which ones are unimportant? The character of life is intimately linked to the character of the evolution of the Galaxy, stars and solar systems and their various alignments. Thus, its survival during the journey around the Galaxy will continue to be determined by changes in these characteristics and alignments.

Although I am painting a picture of a dangerous Galaxy and Solar System, Mother Earth is not a completely innocent bystander in this story; an unfortunate victim in a messy game of galactic and stellar evolution. The threat to life on Earth during the galactic journey also comes from the Earth itself. Inside the center of our planet are the decaying products of its formation; radioactive elements, particularly uranium, thorium and potassium. They heat the planet and help keep the core and some of the mantle, the region just above the core, molten. During the early history of the Earth the molten rock split up into its different constituents, in much the same way that oil and water separate in a glass with less dense oil floating on denser water. Light, porous, silicon-containing rocks solidified and floated over the molten core and mantle, made mainly of iron.

These floating chunks of the crust are the familiar continents on which we live and as currents move around in the molten center of the Earth, the continents drift around. The collision of these continental masses gave birth to some of the mountain ranges of the world. The Himalayan mountain range, home to Mount Everest and the second highest mountain on Earth, K2, is being formed to this day by the collision of two great plates. They are the Indo-Australian plate and the Eurasian plate. The mountains climb into the skies at 5 millimeters a year, a miniscule but measurable amount. I've always been quite enamoured with the idea of climbing Everest and launching a great press campaign when I reach the top, claiming that I am the first person to climb the highest mountain in the World. When I get branded as a liar, I will remind people that the mountain is in fact about a hundredth of a millimeter higher than when it was ascended the day before and therefore I now stand as the first person to reach the summit of the highest mountain on Earth (until of course the next person gets to the top after me). A trite observation, but one that would, probably rather unfairly, make light of our strange human obsession with exploration "firsts".

Some of these plates pass under and over each other. These seams are not perfect and on occasion hot liquid rock from under the crust can come gushing to the surface in the form of volcanoes; some of these happen on land and some happen in the ocean as undersea chains of lava and hot-water vents or "hydrothermal vents" where the continents spread apart. The eruptions are not limited to the edges of the plates, though. Liquid rock can reach the surface as plumes, rising up from the mantle of the Earth and erupting at the surface, sometimes even in the middle of continents.

Of the dangers to life on Earth caused by the Earth itself, these volcanic events are overwhelmingly the most important. Never have hurricanes or giant rainstorms, for instance, been claimed as mechanisms for extinctions of species, but large volcanoes, although still controversial, are a focus of discussions on mechanisms for extinction. Because volcanoes result from the characteristics of the Earth laid down during its accretion in the protoplanetary disk (consider a less volcanically active planet like Mars, which has no present-day plate movements because it is small and the core has essentially solidified), I will consider volcanoes to be part of the selection pressure on life imposed by the cosmic environment. The presence of extinction-causing volcanism is dependent upon whether you happen to evolve on a planet that is large enough or hot enough in its center to support active volcanoes during its history. Furthermore, it has been suggested that asteroid and comet impact events may be a cause of volcanism by punching holes into the molten center of the planet. This idea is widely discredited but, controversial or not, it really is important to talk about volcanoes as part of our 225 million year journey around the Galaxy.

Before I give you any more reasons for sleeplessness, let me at least let you into the revelation that the threats I describe above are quite rare. They happen over geological time periods of millions of years and longer, so you are unlikely to see these in your lifetime.

It might be easy, though, having gained an understanding of the nature of the formation of the Galaxy and the Solar System, to still wonder why the Earth is at risk. Isn't our life a special case of life on a planet? Is it really true, even with all that is known about the violence of the Galaxy, that the Earth is still vulnerable to changes in the cosmic environment? Surely our planet is big enough to be nearly immune from these things? Carl Sagan did great service to our understanding of our place in the Universe in the 1980s. He suggested that the Voyager 1 spacecraft, which was then on its way beyond Jupiter towards the blackness of the outer Solar System, should be pointed back towards Earth and used to take a picture of our planet. On February 14, 1990, a day that should perhaps be celebrated each year as "Pale Blue Dot Day" by leaders of the world to remind us of what we are, the picture was sent to us from 3.7 billion miles away by the spacecraft. You can see this little dot in Plate III. The spacecraft was essentially looking down on us onto the northern hemisphere. In that dot are Europe, North America and much of Asia and Russia.

A small pale blue dot in a middle of a single picture frame. The picture is a remarkable one and one that is relevant to my discussion here because it will give you a better intuitive feel for our galactic journey. Painting a picture in words of the journey around the Galaxy and its possible effects on life and describing where the various perils might come from is one thing, but actually seeing with your own eyes the Earth as a little dot is much more powerful. Now you can imagine this dot oscillating up and down through the galactic plane and traveling around the Galaxy once every 225 million years. Now you can see how this little speck of dust might be influenced by all the things going on around it.

When Sagan first described this picture he brought home to the world the fact that every part of human history that had ever been known occurred on this small dot. As you look at it you can think of every poor person and every rich person that has ever lived. Every ancestor you ever had came from this tiny world. Every terrible crime and extraordinary invention, from the discovery of fire to the invention of spaceflight, has all occurred on this tiny little speck.

Humans are probably too wrapped up in their own affairs to have been influenced by this photograph. Wars didn't stop overnight and peace did not spread across the Earth the day after it was taken. However, likewise, neither the discovery that we are not the center of the Universe nor the theory of evolution made much of an impact on our behavior either, even though both of these advances reduced our significance in the Universe. It is rather a pity because in many ways that grainy photograph, however unremarkable it may look if you haven't read the caption, is perhaps the most remarkable photograph that has ever been taken. In terms of its potential to make us realize our place in the Universe, it could arguably stand alongside evolution and the Copernican revolution as a profound milestone in human self-perception.

I can think of only two other photographs that would be more remarkable than this. The first one would be of our Solar System as a single dot in the middle of a picture frame. Then, not only would we have an insight into the smallness of the Earth, but into the smallness of the Solar System as a whole. We would be getting closer to seeing how we fit into the galactic journey. The second one would be taken hundreds of thousands of light years from the Milky Way, looking back at the Galaxy with a single tiny dot on the edge of a teeming mass of 100 billion dots – our Solar System on its 225 million year journey around the Galaxy. If we had such a photograph it would be the cover of this book. Maybe one day we will.

Sagan went on to say that maybe this picture of this tiny blue dot would remind us that it is unlikely that there is anyone out there to save us from ourselves. I might add that Sagan's analysis applies to life on Earth in general as well as to human society. The broader biological folly of mankind to assume that somehow life is invariant, that humans are the products of a supreme process of biological evolution and that the Earth is some type of protected haven on which life can thrive is a deeply rooted belief. We know that extinction occurs in the history of life on Earth; we see it in the fossil record and we document it in great detail. Still there is a gulf between the scientific reality and our true comprehension of how life fits into this galactic environment; a reason why biologists still today tend to focus on the terrestrial causes of extinctions and the more Earthly selection pressures of Darwinism. Somehow the idea of an earthworm on a roller-coaster ride around the Galaxy just doesn't appeal to a biologist; it just doesn't seem to be what an earthworm is really doing each day, but it surely is.

Before we start to look at how life, and particularly microbes, can cope with these disasters from Earth and space it would be good to know something about the wonderful diversity of microbial life on Earth and the huge variety of nooks and crannies it can hide in. In the chapters that follow we'll visit each one of these disasters. We'll visit the Earth in the midst of a near-by supernova explosion and we'll go to a different time just after the Earth has been hit by an asteroid. We'll visit the Earth during a massive volcanic eruption. As we hold on for our roller-coaster ride through these disasters we'll go and visit some of these habitats and see how life is faring and what makes it through.

We'll begin to ask our question – what is the effect of the cosmic environment on life? During a journey once around the Galaxy, what perils can life expect to face and how does the microbial world fare? Now I have introduced you to our origins and where the Earth came from, let's go on a tour of the microbial world to see what's living here.

3   The microbial menagerie

There is a television show that aired in the USA and Europe called *Lifestyles of the Rich and Famous*. In it we are shown the homes, lives and habits of some of the World's stars. We tour the embellished and flamboyant lifestyles of those who have made the big time. The purpose of the show, apart from of course the fascination of seeing other lives at work, is to get an insight into a lifestyle few of us experience. A show that I always wanted to host would be called *Lifestyles of the Small and Lowly*, and would be about the homes, lives and habits of microbes. We would visit some of the World's most extreme environments and see how they thrive and live in the hottest, coldest and most poisonous environments. My show would be aired after the *Rich and Famous* and then at the end of the show I would hold a national TV ballot to find out which the public found more impressive, rock stars or microbes.

The impressiveness of microbes lies in their pervasiveness and their versatility. They come in all shapes and sizes and in this book I am using the term "microbe" quite loosely to cover any organism that essentially lives as a single cell, unlike you and I, who are made up of trillions of cells. As we'll soon see, different microbes eat different things and they live in very different environments. They range from the tiny *Escherichia coli* (often just called *E. coli*) that inhabits your gut, a rod-shaped microbe just a thousandth of a millimeter around its waist (Plate IV) to the beautiful forams, a group of microbes that live, among other places, in the oceans. With ornate shells made of calcium compounds that they scavenge from the ocean waters, they can be up to a millimeter in length (Figure 4). Some of these microbes are very ancient and form the lowest branches of what is colloquially called "the tree of life". Others, like our humble gut microbes, may

FIGURE 4. Foraminifera, marine microbes that eat other microbes and live in sediments, are some of the largest microbes to be found. Their size is caused by the shells or tests they excrete around themselves, the purpose of which is still a matter of some debate. These tests, made up of sand, calcite, organic matter or a variety of other materials depending on the species, are up to a millimeter across, a thousand times larger than the *E. coli* in Plate IV. (Image, Wim van Egmond.)

be more recent appearances on the stage of life. Before we have a look at life on Earth today, let's take a brief journey back to life on the early Earth, because this will give us an insight into where this remarkable diversity of microbial life first came from.

Just under four billion years ago, when the Earth had cooled sufficiently out of the cloud discussed in Chapter 1, life arrived on the scene, preserved in the fossil record. (As with the origin of the Universe let's not get into a discussion about how the first single-celled microbe came into existence. This branch of science, which covers the chemical origin of life, is a truly fascinating one, but my discussion here focuses on what happens to life once it does arrive, not on the complexities of how it might have got there.) During this time some of this life is preserved as "microbial mats". The first microbial mats,

MORE ABOUT THE TREE OF LIFE

Microbes are related to each other on what is often called the "tree of life". The tree of life defines the distance between different domains, kingdoms and families – all the way down to the different species of microbes. The branches on the tree represent the time when these different organisms split into different groupings. It is based on the differences in 16s RNA, a strand of the genetic machinery of the cell. Microbes are found in all the branches of the tree. They include the bacteria (including our gut bacterium E. coli), the archaea (a branch of microbes that include some of the most primitive families such as the deep-sea varieties or varieties that live at extremely high temperatures) and the eukarya, a branch which includes ourselves and all other animal life. The eukarya also include a huge diversity of microbes such as algae and fungi (Figure 5).

so called because, well yes, they look like doormats, can be found in the fossils as layered assemblages of microbes one on top of the other like a sandwich with a mineral glue in between. Modern representatives of these types of structures can be found in some shoreline areas of the world, including Western Australia and California.

There are other clues to the presence of life early in the history of Earth that are perhaps more subtle than the fossils of mats (and some believe the mat fossils are quite controversial). Some elements, like sulfur or carbon, have what are called "isotopes". An element is said to have isotopes when it has two different forms that have a different number of neutrons (a subatomic particle). The effect of these different numbers of neutrons is that the two different isotopes are chemically similar, but they can be physically different. The most important difference is that the one with fewer neutrons is less heavy. The sulfur isotope sulfur-32, for example, is slightly lighter than sulfur-34. There are a variety of microbes that like to use sulfur and its various compounds as a form of energy. One type are called sulfate-reducing microbes because they can use the compound sulfate (which is sulfur bound up with oxygen atoms and very common in the oceans), to get the energy reserves they need. Now it turns out that the microbes

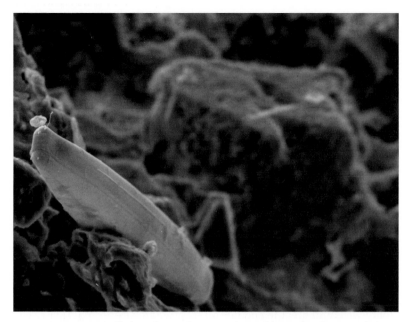

FIGURE 5. All shored up. Like a run-aground boat, this diatom inhabits the bottom of a lake in the Canadian High Arctic at 75° N. These microbes, that have shells or frustules made from opal that they scavenge from water, are about a twentieth of a millimeter long and inhabit the sediments as well as the water column of the lakes. This one, photographed using a scanning electron microscope, is attached to a small rock particle from the lake on Devon Island. (Photo, Charles Cockell.)

have a preference for light isotopes, so that they use more sulfate that has sulfur-32 in it than the sulfur-34. In a rock made up of fossils of these microbes there is an unusual amount of this light sulfur compared to the heavier one. Geologists can use instrumentation to examine the quantities of these different isotopes as a method of deducing the presence of life.

So although we can't necessarily see the fossil remains of life, the signs of their activity are there to find. These sulfur signs of life have been found in rocks dating to the early Earth. In fact, some deposits seem to have been formed almost exclusively by the activities of microbes, showing that they could have been a very important component of early life on Earth. Looking for signs of life like this can be done with other elements as well. Microbes that use sunlight as an energy source take up the gas carbon dioxide. They prefer to use carbon dioxide with a light carbon-12 form of carbon compared to the heavier carbon-13. Like the sulfur compounds, an excess of carbon-12 in a rock can tell us that life that used sunlight as a form of energy has been at work. So fossil evidence and chemical evidence can be used side by side to begin to build up a picture of life on early Earth.

Environmental conditions on the Earth four billion years ago were very different from the ones we are familiar with today. In this period of Earth history, called the Archean, the atmosphere had much higher concentrations of the greenhouse gas carbon dioxide, probably about two thousand times higher than they are today and the concentrations of oxygen were probably very small. So it was not a place where we could breathe and nor, in fact, where any animal that needs oxygen was able to survive. Yet the fossil and chemical evidence shows that microbes could exist and they did so in abundance.

At this time the first continents began to form and the majority of life was probably confined to the oceans. Scientists don't know much about the temperatures on early Earth, but they may not have

MORE ABOUT GEOLOGICAL TIME PERIODS

Earth history is split into a number of eons. The Archean eon runs from about 4 billion years ago to 2.5 billion years ago. The Proterozoic eon runs from 2.5 billion to 544 million years ago and the Phanerozoic eon runs to the present time. Some people, depending on who you talk to, also refer to these as "eras", although generally eras are used for the next subdivisions of geological time that are themselves split into periods. The Hadean that runs from about 4.5 to 4.0 billion years, constitutes the period of history during which the Earth was formed and coalesced. It is the period in which it is supposed that the origins of life occurred. Humans appeared only during the last 2 million years of this incredible 4.5 billion year history of Earth.

been radically different than today. Gases that were being produced by volcanoes, like hydrogen and hydrogen sulfide were mopping up the oxygen. Because there was no oxygen in the atmosphere, an ozone layer, which protects us from many of the harmful effects of ultraviolet radiation and which forms from oxygen, could not exist as it does today. If you could take a time machine and travel back to Archean Earth, you would get a suntan from the ultraviolet radiation about a thousand times faster than you would do on Earth today – in other words you would be burnt very quickly. There could have been other gases in the atmosphere that shielded from ultraviolet radiation. If the gas methane, a flammable gas associated with bogs, had been produced in the early atmosphere, then it could have reacted to form a smog in the upper atmosphere that might have created a shield for life.

Even if there was no shield and the ultraviolet radiation was much higher than today, there are ways microbes could have survived. They could have hidden. By hiding under the top layers of microbial

MORE ABOUT ULTRAVIOLET RADIATION

Ultraviolet radiation, often called "UV radiation" for short, is the part of the spectrum beyond violet. Humans can't see it with their eyes, but some insects, lizards and fish have UV vision and they can see this color of light. Ultraviolet radiation has a lot of energy in it, which is why it is dangerous to life. You'll know how damaging it can be if you have sat out in the Sun for too long and been sunburnt. Most of the UV radiation produced by the Sun is screened by the ozone layer in the upper atmosphere, which is why people are concerned about the "ozone hole" over the Antarctic, caused by pollution, which increases the amount of UV radiation that reaches the ground. Ultraviolet radiation is particularly damaging to the genetic material, DNA, which contains the information for life. Because DNA is involved in passing information from one generation to the next, if it is damaged it can cause problems, particularly to some microbes that may only have one copy of this information.

mats, ultraviolet radiation can be almost completely avoided. Some microbes are very good at repairing ultraviolet radiation damage as well. There is a microbe called *Deinococcus radiodurans* (pronounced Dine-o-cocus ray-dee-o-dur-ans), of which we will learn more later, that is supremely successful at repairing breaks in its genetic material caused by radiation.

People that study the Archean period should really consider second jobs as forensic scientists, because their skills are honed in using small pieces of evidence to build up a picture of what the conditions of early Earth were like. Most are agreed on the fact that, despite the hugely different conditions on the planet compared to today, Earth was a thriving planet of life and that these microbial communities may have been quite diverse with complicated interactions. Perhaps they were sufficiently complex that we would recognize them even today as complicated microbial ecosystems.

For about three billion years microbes dominated the Earth. You might ask why the Earth remained a microbially dominated planet for so long and why it took so long for bigger animals, like us, to arrive. "Why did only microbes live on the Earth for the first three billion years of life's history and animals only come into existence 600 million years ago? What took so long?", are not uncommon questions. This is a very human view of the world. Microbes are a supreme evolutionary achievement. They are small and so they can inhabit the tiniest spaces in which water and energy are available. Some can harness the most basic form of energy, chemical energy, and many can go dormant for long periods of time, possibly millions of years, waiting until better conditions arrive. Many can divide into two in less than twenty minutes. More accurate questions are, "What went wrong 600 million years ago? Why did microbes suddenly regress into large lumbering, specialist animals prone to extinction?"

About two billion years ago microbes became responsible for the largest pollution event in the history of the Earth. After mastering the means to get energy from sunlight, photosynthesis, they started producing oxygen, which is a waste product of photosynthesis. The biological oxygen factory was so continuous and extensive in its production that once the gases from volcanoes had all been mopped up, the atmosphere became filled with oxygen. This is what went wrong.

Oxygen is an explosive gas. By burning carbon compounds with oxygen, enormous energy resources can be released in short periods of time (indeed, at high enough concentrations of oxygen, perhaps above 40%, some organic materials like wood or paper can spontaneously combust). It was this energetic possibility that soon came face to face with evolutionary innovation. Such a possibility for energy production could not go untouched by the continuous process of mutation and selection. This oxygen would provide the fuel for a new intense form of biological energy production – respiration, manifested in animals as breathing.

This new form of energy production allowed for bigger organisms. It provided energy that could support running, flying, jumping. Leopards, kangaroos and birds were now on the cards. But respiration is short-sighted consumerism. By releasing energy on such scales organisms can be bigger and better than ever before, but the "better" part of it is the error in our perception. Size makes it difficult to escape from asteroid and comet impacts, supernova explosions and volcanoes. Being large means that you need lots of carbon to burn as well. Because photosynthetic organisms produce by far the greatest biomass on Earth, access to sufficient carbon to burn means eating plants, photosynthesizing microbes or other animals that themselves eat organisms that photosynthesize. And so by locking into oxygen, large animals had inadvertently locked themselves into photosynthesis. This had two fundamental consequences. Firstly, most of them need to live near the surface of the Earth to get access to their food, because that's where photosynthesizing organisms live. They immediately relinquish the vast diversity of subsurface habitats that are open to microbes and this has important consequences for their survival during the galactic journey as we will see later. Secondly, they are vulnerable to starvation if the light goes out. Large animals had put themselves at risk of complete extinction if there was a sudden development of hostile conditions on the surface of the Earth and particularly if there were darkened atmospheric conditions that block out the light for photosynthesis.

The pollution of the Earth's atmosphere with oxygen had therefore spawned remarkable organisms that were unconscious of their own vulnerability in the context of the long-term statistical certainty of catastrophic change. Only one organism would emerge that would have the intelligence to see this vulnerability and perhaps be able to act on it; to seek out asteroids and comets, to observe supernovas and their frequency in the Galaxy and to study volcanoes and how they occur – and that is ourselves. Whether that intelligence will be enough to save us from the perils of our galactic journey has still to be determined.

During their early evolution microbes made great innovations in the use of energy supplies. All microbes need energy; they need energy to divide, to repair their DNA and to take care of the other housekeeping functions in the cells. Sunlight is a very good way to get energy, as your houseplants will testify. However, wouldn't life on Earth be limited if microbes had to rely on being exposed to sunlight all the time to get their energy? They would be completely restricted to the exposed surface of the Earth. Many microbes are covered by soils, they get carried underground in water or they live in rocks. To inhabit these types of environments you need an alternative source of energy. The innovations that would allow microbes to go underground and under cover to exploit new sources of energy would turn out to provide some lucky means of escape from the threats imposed upon life as it traveled on the galactic journey.

There is another way to make energy that doesn't use sunlight and on the early Earth microbes were busy evolving to be able to do this. Instead of sunlight, energy in chemical compounds can be used to drive life. For example, let's look at the compound iron, sometimes more widely known when it becomes rust in your car or your garden equipment. Different forms of iron have different numbers of subatomic particles called electrons. The type of iron called ferrous iron has one more electron that the type called ferric iron. The different numbers of electrons represent a form of energy. There is a compound in microbes called ATP or adenosine triphosphate. This is a high-energy compound that microbes can use to fuel all the energy-intensive needs they have in their day-to-day living. Any chemical reactions that can release electrons can be used by microbes to make ATP. Because there is one electron difference between ferric and ferrous iron, by converting one form to another, electrons can be released by the humble microbe to manufacture the energy-full compound ATP. This is a cunning way to get energy from simple elements in the Earth's crust. This is the basis of what is called "chemosynthesis", the use of chemical reactions to get energy. Scientists believe that this method of getting energy evolved in the earliest microbes on Earth.

Getting energy from basic chemical reactions is an interesting system because it allows microbes to completely escape the need for sunlight. If a microbe with the right systems inside it can find some iron in its environment then it can take it into the cell and remove an electron. The electron can now be used to do some useful work. There are a plethora of different reactions like this that microbes can use. They can use different compounds of sulfur, nitrogen and many different metals like manganese. Different microbes, by living in the same environment, can use each other's wastes to drive their energy needs. For example, a microbe that uses the compound sulfate can produce sulfide as a waste product. The sulfide reacts with ferric compounds and the ferric iron turns into ferrous iron. The ferrous iron is then food for yet another microbe and so on.

A most remarkable place to see these amazing transformations happening is on the wreck of the RMS *Titanic*, four kilometers deep in the North Atlantic. The wreck is covered in over 650 tonnes of "rusticles", giant icicle-like formations made up of microbes. No one knows precisely how rusticles work, but some insights into the microbiology of these features are beginning to emerge from deep-sea dives to the wreck. As scientists learn more about them it is becoming obvious that they are more that just large pieces of rusting *Titanic*. There are iron-using microbes in the formations and sulfate-reducing microbes. One theory is that sulfate-reducing microbes in the rusticles are changing the sulfate in seawater into its more reduced form, sulphide. The sulfide is turning the ferric iron in the steel plates of the *Titanic* into ferrous iron and the ferrous iron is being used by other microbes. These colonies are destroying the ship at an incredible rate. Within fifty years the ship may have fallen apart, entirely because of the microbes on its hull.

As well as degrading the *Titanic*, these types of reactions allow microbes to exploit the elements found naturally in the Earth's crust (the elements I told you about earlier that originally came from the supernovas). Iron is the fourth most abundant element in the Earth's crust and so it doesn't take a great leap of imagination to see what would have happened in the earliest stages of evolution to a microbe that evolved the ability to metabolize iron. It would have been faced with a huge food resource that was inaccessible to everything else around it. By exploiting iron it could have multiplied with virtually no competition (although it might be limited by a lack of other nutrients). This argument follows for many other elements. In any environment where an element is to be found that is not in use by a microbe and there are microbes competing for energy supplies, there would be some advantage to having a mutation that could get energy from that element. Slowly, over the first millions of years of the Earth's history microbes exploited almost every chemical reaction that could yield energy, leading to the diversity that is seen today. Once they had learnt to harness these energy resources then they were set to take over almost every habitat on Earth where there was also liquid water.

My fascination with microbes that use chemical reactions is because the food they eat is the raw stuff of planets. Rocks contain elements and elements are energy. You can't get more basic than that. And in some ways it is these reactions that make the possibility of life on other planets so compelling. If life required some complex series of molecules as an energy source then it might be difficult for us to imagine life elsewhere, such as on Mars. However, the recognition that life has successfully been able to harness even the most basic chemical reactions to generate energy gives us some indication that it could happen on a place like Mars. But it also tells us something more compelling, that I'll return to again later. If microbes can eat the most basic elements and use them to make energy, then if they leave the Earth and go somewhere else where these elements are to be found, then they could potentially arrive in a place with plenty of energy reserves. And so the simplicity of this way of life is not only the key to the conquest of all habitats on Earth. Speculatively, it may also be the key to the exploitation of habitats on other planets that microbes happen to find themselves transported to on pieces of rock thrown out into space by asteroid and comet impacts.

I could end my discussion of how microbes manage to use the incredible diversity of energy sources on Earth in different ways now. However, before I do that, it would be impossible for me to pass over the most gruesome form of getting energy – predation – or eating your neighbor. Not a trite diversion from our discussion, but a mode of existence that might have some significance for surviving the galactic journey.

Microbes themselves are a good energy source, and certainly if you are a microbe they are. By definition they contain all the elements that you need to survive, so by simply eating your neighbor you have an instant packaged meal with all the elements in the right combinations and quantities to supply your needs. The microbe *Bdellovibrio* can penetrate the cell membrane of other microbes and live inside its host. Inside its victim it slowly eats up the insides until the cell contents are exhausted and the prey breaks up, releasing the offspring of the invading microbe. More direct predatory behavior is observed in the myxobacteria. These microbes release enzymes outside the cell, "exoenzymes", that bathe their prey and break up the prey bacteria. The myxobacteria can then suck up the disrupted cell material and digest it.

Most people think of predation as actually eating your prey; like the classic chase between a lion and a gazelle. There is a group of microbes known as the dinoflagellates, that are able to swim up to their prey. Using small tails or "flagella" to move about, they are true examples of predation as it might be classically seen. These microbes can use sunlight to get their energy when there is sunlight around, but as soon as the light fails, they start to eat their neighbors. They swim up to an unsuspecting microbe (preferably one that can't swim so it is powerless to escape) and then they envelope the prey. The tiny unsuspecting microbe is then completely engulfed and digested. This system of living has great merits. It means that the dinoflagellates aren't constantly relying on sunlight for energy. If they get caught up in a current or drift into a river or pond where mud or other contamination cuts out the light, they are not necessarily going to starve to death. Provided there are other microbes around them, they can live by eating them. By eating other microbes they can achieve incredible growth and reproductive rates. Many microbial ecosystems from small ponds to the open oceans have high numbers of the predatory microbes.

This mode of existence, of photosynthesis interspersed by predation, is called "mixotrophy", because microbes can use mixed energy sources. The mixotrophic way of life is one of great interest in terms of catastrophism because, as we will see later, one of the suggested effects of large volcanic eruptions and asteroid or comet impacts is to throw up dust and soot into the atmosphere, blocking out sunlight. So a microbe that can use sunlight, but resort to eating other microbes when the light goes out, might be well suited to the unpredictable changes of our galactic journey.

Gruesome stories of microbes eating other microbes are not just frivolous. They illustrate the adaptations that the microbial world has mastered in the struggle for energy to ensure the continuation of their genetic material from one generation to the next.

Getting food when there are plenty of energy sources or neighbors around is a nice life, but no one can be this certain that the next meal will turn up, as you and I have probably found out on occasion. Microbes need to be able to deal with a little starvation now and then. In times of austerity, microbes can resort to a suite of adaptations that allow them to survive. These systems could help them to avoid starving to death when there is a paucity of food in their habitat or, equally, when the global biosphere is under stress during a calamity thrown at the Earth during the course of its voyage around the Galaxy. One way they can respond is called "the stringent response", where a number of energy-draining reactions, like the production of proteins and genetic material, can be shut down or slowed down, thus preserving energy. Another method is to switch food resources entirely. There is a gene in some microbes called the "nac" gene. This gene allows the microbes to switch from getting nitrogen that they need for their growth from their normal sources, like nitrates in the soil, to a whole range of other organic molecules that happen to have some nitrogen in them. The "pho" system works in the same way for phosphate, an important element for making enzymes, sugars and DNA. So as well as just cutting back and not consuming so much, there is the option to try and find new ways of getting the resources they need.

There is another way to survive periods of starvation that is more dramatic. That is to shut down altogether, to go into a state of complete dormancy. Most large animals that hibernate during periods of austerity can only do it for short periods of time. The polar bear, for example, will run out of fuel reserves if it doesn't come out of its lair to hunt for seals in the spring. Hedgehogs must come out of hibernation and look for food in spring, they would quickly starve if they tried to survive a second summer and winter without hunting. The reason for this is that mammals are complex animals. Even when they are hibernating, they still need energy to keep their hearts going, to keep their body temperatures up and their immune system active. They never completely shut down. Microbes, on the other hand, really can shut down. Some can form spores, which have a coat of material that is highly resistant to environmental damage.

These spore formers are believed to be able to survive for millions of years. The species *Bacillus subtilis*, a microbe normally found in soils, was sent into Earth orbit on board a rocket and survived as spores for six years in the vacuum and the cold of space before being returned to Earth, protected by its spore coat. They have no need to keep any type of metabolism going. By completely drying up they can shut down all processes that would otherwise drain energy. Provided that during their period of dormancy their DNA is not so damaged by the environment that when they come to life again they can't read their genes, then they will continue to be able to grow. By using starvation strategies microbes can survive periods of difficulty, but by forming spores they can just pull out of the starvation race altogether, waiting until conditions improve.

Spores have a great deal of significance during the galactic journey. In any environment, even the soil in your garden, there will be microbes in various states of hibernation. A spore a kilometer under the ground that happened to get washed there in groundwater could survive a calamitous asteroid or comet impact event quite fortuitously. A thousand years after the disaster another underground stream could return it to the surface where it could begin to grow and multiply. If you consider that these spores and other resting states are scattered all over the world on the land and in the oceans, then in some ways the microbial world is ready for disaster. It has, through the vagaries of everyday living on Earth, already set itself up with a global repository of generations of microbes that are ready to emerge at some undefined point in the future after other organisms have become extinct. The microbial world is well prepared for the galactic journey.

Large mammals and things we can see with our naked eyes bedazzle us. This is partly because we can observe them without much effort and partly because we see ourselves in them. Cuddly polar bears, baby seals and puppy dogs all appeal to the human mind because we think of them as vulnerable, like babies. Campaigns for endangered species almost exclusively focus on them. When did you last see a

T-shirt proclaiming "Save the fungal spore!" or "Save the endangered deep-subsurface microbe!"? However, over 80% of the Earth's biomass today is locked up in microbes. Almost all of the large-scale biological cycles of the Earth, the exchange of carbon, nitrogen and sulfur, owe their efficiency and scale to the microbial world. For example, each year microbes extract 140 million tonnes of nitrogen from the atmosphere and deposit it as biologically available nitrogen compounds into the soil. This nitrogen fixation is essential for plants and thus for higher animals that need these compounds to grow, including ourselves. Without microbes, long ago there would have been a nitrogen crisis on the surface of the Earth. Large animals are a mere veneer on the microbial world and they are completely dependent upon it.

It is clear that the despite the emergence of animals, humans live on a microbially dominated planet and the habitats that microbes live in are as diverse and extraordinary as they probably were on the early Earth. So just how pervasive are they today?

The shaking is quite terrible, but that's what you would expect from an imaginary human-operated subterranean drill. I do apologize, but I have volunteered you as my companion on some expeditions to Earth's extreme environments. We're going to go and have a look at the microbes that live in these places and I think you'll be fascinated, which is why I volunteered you. The giant drill-piece roars, jumps and vibrates around. Through a small porthole built into the side we can see out into the brown soil that is now filling our view and blocking out the sunlight above us. The soil leaps around in front of our eyes. We are on a journey into the bowels of the Earth. We have not come on a sightseeing journey. We have a microbiology laboratory with us and we can take samples from outside the drill when we want to. We are traveling fifty meters an hour, a pretty remarkable feat of engineering. And after several hours of being thrown about we are at our first destination, half a kilometer under the ground.

At this depth in the ground the microbes are abundant. There are up to ten million individual microbes per cubic centimeter. Many different types of microbe can be found in this region of the Earth's subsurface. The iron-and sulfur-using microbes I talked about ear-lier are in the soil and they find plenty of minerals to give them energy. Cracks and channels in the ground carry groundwater from rivers and rain deep into the ground and this provides them with the water they need. Estimates of how many microbes live at a depth of half a kilometer vary, but some think that the total mass may be up to 10% of all the biomass on the surface of the Earth – a staggering amount of life. At this depth a standard garden bucket of soil would contain more microbes than there are stars in our Galaxy. The subsurface of the Earth is an unseen repository of life hiding from the unpredictabilities of the surface world.

Satisfied that at a depth of half a kilometer the Earth is seething with life we can continue on our journey, spending another few hours of bone-jarring discomfort to reach a depth of one kilometer, just under a mile deep in the ground. At these depths life continues to thrive, but the types of microbes can be very different from the ones higher up, particularly if they live in rocks. Instead of eating minerals, some bacteria can use the gases hydrogen and carbon dioxide to thrive. The hydrogen, which comes from reactions of water with the rocks, is an energy source and the carbon dioxide provides the carbon they need to build basic molecules. These energy supplies are enough to keep them going and a huge biomass of microbial life continues to thrive even at these great depths. Deep in the Earth the temperatures are higher than at the surface because we are receiving more heat from the molten center of the Earth. We need to switch on our air-conditioning system so that we can stay comfortable. If we didn't have a bit of cooling we would soon heat up to about 40 °C.

As we go deeper into the Earth, the high temperatures begin to become a problem for life. The temperature of the Earth increases by twenty to thirty degrees centigrade each kilometer we go down, and once we get to a depth where temperatures are above about 110 °C, life begins to disappear completely. We find that this depth is about five to six kilometers. At these depths and just above, where tempera-tures remain cool enough, microbes can still use hydrogen and carbon dioxide to get their energy and carbon needs and there is enough water percolating down for them to thrive. Below these depths we can still find the remains of dead microbes. These "biomarkers" have been found at depths of greater than six kilometers.

We can perform a simple calculation to demonstrate just what mass of microbes there might be in the subsurface of the Earth. If we assume that rocks have about a fiftieth of their space available for microbes to live in and about a hundredth of this available space is taken up by microbes that have a density similar to water (remember that all organisms are about three-quarters water), then to a depth of six kilometers the entire subsurface of the Earth could support 460 trillion tonnes of microbes. Of course there will be places where there is little life at all, so this calculation is probably not very accurate, but it gives you some idea of the possible vastness of the subsurface microbial world. This is no ephemeral habitat for a few haphazard bugs that get washed into the ground. It is an unseen biosphere of its own, with a diversity and productivity that rivals many places on the surface of the Earth. It is as much a part of the biological world as a rainforest or a savannah, but it is one dominated by microbes.

The drill expedition was an experience, but nothing to what we are about to do now. You've got the window view this time and it doesn't look out into the brown soil of a subterranean expedition, instead it looks into a clear blue sky. Strapped into the Space Shuttle, we have been assigned our task. At a height of seventy kilometers we are to activate an air sample collector outside the spacecraft. Our purpose for this mission is to retrieve air samples halfway into space. Then we will land the Shuttle. We will look at our samples under the microscope to see what we've got. The idea of the mission is to see how high microbes can go.

As the Shuttle climbs into the sky the numbers of microbes in the atmosphere slowly decline. The higher we go, the more the pressures and temperatures drop and the greater the ultraviolet radiation becomes. At a height of about two or three kilometers the temperatures drop below freezing and any microbes found here are in a state of dormancy, frozen onto bits of dust or ice. Their survival at these heights and much higher is undoubted. During the 1970s Russian scientists using collectors on the top of rockets found spores of fungi floating around seventy kilometers above the ground. These fungal spores were found to grow when they were returned to laboratories and placed in dishes with nutrients. On our mission, as the Shuttle touches the envelope of space, we will almost certainly return to Earth with samples of microbes. The ways in which they reach these heights are a matter of some speculation. Hurricanes, dust devils, tornados, volcanoes; there are a number of ways that microbes can be picked up from the surface of the Earth and catapulted into the high atmosphere. And once they are up in the sky they will be carried along by winds until they get caught in downdraughts and float back to the Earth. An exchange of microbes between the atmosphere and the surface of the Earth is happening all the time.

There is a big difference between the subsurface biosphere and the high-altitude one though. Many of the microbes floating in the sky are not growing, they are more on a trip from one place to another, whereas underground microbes are often actively growing. The sky isn't a very good place to grow because liquid water is scarce. Any water that microbes have around them is soon dried away or frozen. Unless they happen to be stuck to a particularly rich dust or soil particle, the energy sources and elements needed for their growth are not available either. In some very cloud-ridden places on Earth, such as in Chile, microbes might actually be able to grow if the clouds stay around for long enough. Clouds could provide water aerosols, tiny drops of liquid water, in which microbes would grow. Microbes have already been enticed to grow in artificial mists in the laboratory, so there is some reason to suspect that clouds and fog can be a home to microbes. Generally the sky is a rather transitory habitat though, as you will know if you have watched a cloud passing by and seen how much it can change in just minutes.

If microbes can travel around the Earth then any habitat with some water might be colonized provided that the right microbe drops in. You can think of the sky as a vast reservoir of microbes; some like it cold, some like it hot, some like acid, some like alkaline. They are thrown up into the sky from every different habitat on Earth and they are raining down on other habitats, the ones that survive are the ones that happen to drop into the right habitat. This is probably quite simplistic, and many microbiologists would argue fervently against this view of the microbial world, but it may help explain why, if a new volcanic hot spring sprouts up next to a new volcano, heat-loving microorganisms will very quickly colonize it (Figure 6). It might also explain why diseases can be carried around the world. Proving these relationships is very tricky, but of the fact that the sky is filled with microbes traveling around the world, there is no doubt.

We've been into the ground and up into the skies – what about the oceans? The Earth is three-quarters oceans and so the life that inhabits the oceans is of great importance to the biosphere. As with the

FIGURE 6. Hot springs in Yellowstone National Park, USA are home to a rich diversity of microbial life. Depending on the temperature, different microbes are able to grow, causing gradients of microbial growth as the water spews from the geothermal ponds and cools along its path. (Image, Thomas Broch, University of Wisconsin-Madison.)

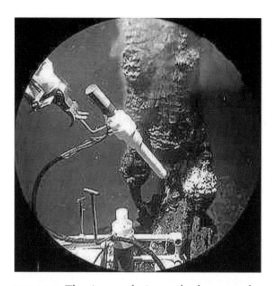

FIGURE 7. The sipper, a device on the deep-sea submersible, ALVIN, collects water samples from close to a hydrothermal vent to study the chemistry of one of the towers of microbial communities that grow where the continental plates rift apart. (Credit: University of Delaware Graduate College of Marine Studies/Woods Hole Oceanographic Institution.)

deep subsurface of the continents, the deep oceans are also home to an astonishing diversity of microbes. One way to reach these deep-sea communities is to use the submersible ALVIN (Figure 7), introduced in 1964. The submersible, which belongs to the Woods Hole Oceanographic Institution in the USA, uses a spherical hull made of the strong metal, titanium, to cope with the pressures encountered when traveling to great depths. Every ten meters you go down in the water is equivalent to an extra atmosphere of pressure pressing onto you, so you can see that at depths of two and a half kilometers, the hull must be capable of withstanding formidable pressures.

Our destination is the hot sea vents where the continents spread apart and the hot gases from the core and mantle of the Earth come bubbling to the surface. The vents of hot water and chemicals that spew from these actively rifting areas in the center of our oceans have for a long time been regions of great interest to biologists. They connect the oceans to the deep subsurface of the Earth. What biologists found at the ridges in the 1970s continues to stagger them today. The chimney-like formations of rock (shown in Figure 7) that form around the vents are complex towers of microbial communities. Like the rusticles of the *Titanic*, many chemical cycles, helped along by microbes, are occurring over very small distances in the towers. Over a distance of just a millimeter you may have patches of microbes turning sulfate from seawater into sulfide and sulfide itself being eaten by a microbe turning it into sulfur. The diversity of these processes is made more remarkable by the physical conditions in which they live.

At a typical vent the pressure is about 250 atmospheres. At these pressures water no longer boils at 100 °C, but at higher temperatures. You can have liquid water as hot as 150 °C that is still not boiling. Microbes have evolutionarily responded to this simple physical fact by developing methods to grow in this super-hot water. The proteins that make up the microbes fold up in ways that help them to be stable at high temperatures. The membranes of the cells are made with lipids that are very fluid, allowing the membrane to move around at high temperatures. So adaptations within the cell allow it to grow in this extreme environment. As I write this book the upper temperature limit for life is set at 113 °C, which is the maximum growth temperature for the microbe, *Pyrococcus fumarii* (Figure 8), found in these deep sea vents.

The upper temperature for life may well go beyond this, but it is unlikely to go much higher. At about 250 °C the energy imparted to molecules becomes so high that most molecules that are important to cells disintegrate in a second or less. It would require phenomenal repair processes and adaptations to keep molecules together at these temperatures and, even if it were possible, the repair processes themselves would be liable to break down. It would appear that there is actually a physically defined upper temperature limit for life. However, 113 °C is quite impressive. Growing at high temperatures doesn't just expand the number of habitats microbes can live in because they can tolerate the heat. Life at high temperatures means life at lower depths

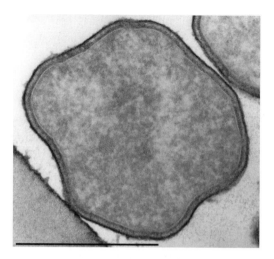

FIGURE 8. A microscope cross-section of *Pyrococcus fumarii*, which currently holds the record for the upper temperature limit for life at 113 °C. The bar is one thousandth of a millimeter. At the high pressures of the deep ocean water boils at well above 100 °C. Life can adapt to these high-temperature environments. (Image, Professor Dr Karl Stetter and Dr R. Rachel, University of Regensburg, Germany.)

in the land and oceans and this means life at depths less disturbed by changes at the surface, which is important for the survival during the galactic journey. The higher the temperature of life, the deeper the refuge, the less prone microbes become to extinction.

Living in these incredible extremes may have some advantages for surviving other problems encountered during the galactic journey. Some of the microbes that live in hydrothermal vents also happen to be quite resistant to radiation. Like a Rudyard Kipling *Just So Story* we might wonder how on Earth these microbes got their radiation resistance when there doesn't seem to be much radiation near these vents. The reason is probably that high temperatures can damage the genetic material, DNA. Radiation damages DNA as well, so any microbe that is able to mend its DNA because of the high temperatures experienced living near a vent will also happen to be quite good at surviving high radiation doses. In other words, because heat-loving microbes have found ways to cope with damage caused by high temperatures, they might just happen to be very good at surviving the high radiation levels that some have suggested could come from a supernova explosion.

Imagine one day you buy a giant plastic tub to grow plants in. You keep your tub carefully for a few years and enjoy the fruits of your gardening. Then suddenly a storm ravages your town and your house is flooded. Stranded in a lake, you use your plastic plant tub as a boat to paddle to safety and save some of your precious belongings. The tub you bought for a completely separate reason turns out to be of use in a serious and quite unexpected situation. During that unpredicted flood, you are thankful that you happened to have such an interest in gardening.

Survival of rare and unusual events will often be survival of the luckiest. The adaptations that microbes have to get into the air, to the depths of the oceans, to the deepest ice cores and to the hottest vents in the oceans may well turn out to provide some lucky resistance to the perils of the galactic journey.

The microbes do more than live on their own around the vents. They also live inside the stomachs of tubeworms. Like a garden of tentacles, hundreds of these worms swarm around the vent, enjoying the heat and microbial growth that keeps them alive. The worms, called *Riftia pachyptila*, contain within them a whole consortium of microbes. These lodgers, through the process of eating carbon dioxide, supply the tubeworms with sugars, which they can burn as an energy supply. In these worms we have an extraordinary example of how microbes, living in extreme environments, can not only survive, but can also make complex life possible.

The hydrothermal vents are in deep water, but not the deepest. That accolade goes to the Mariana Trench eleven kilometers deep in the Pacific Ocean. Much of the life in the Mariana Trench is not near hot vents and so it lives at cool temperatures of about 10 °C or less, but what interested microbiologists about these deep-sea microbes far from the vents was the enormous pressures they could tolerate. At the depth of the Mariana trench the pressure is equivalent to one thousand atmospheres, in other words a thousand times the pressure we feel on us every day. These pressures are a challenge, although they are not insurmountable. Like all life at hot temperatures, microbes have a number of important adaptations to help them survive. Highly fluid membranes and pumps in their membranes that allow them to collect food and expel waste are just some of the necessary adaptations. Studying these microbes is a challenge, though, because it is difficult to build equipment in the laboratory that will stand these pressures without exploding. It is also important to retrieve these microbes from the bottom of the ocean without changing their pressures too abruptly.

The amazing reason for needing high-pressure equipment to collect these microbes is that some of them found at this depth have evolved to *need* high pressures for growth. These "barophiles" (literally, "pressure lovers") often cannot grow at the pressures you and I experience. If they are brought to the surface of the water, not only do they stop growing, but they can also die. Some speculate that this might be because some of the proteins in the microbes fold together in just the right way at high pressures. This folding is disrupted when the pressure is lowered. Very few laboratory groups around the world study deep-ocean microbes, because of the specialist equipment that is needed. To make these microbes extinct one would need a catastrophic calamity that would reach down as far as the Mariana Trench. As we will see later, only the most mighty asteroid or comet impacts could have a chance of killing the barophiles in the Mariana Trench.

I want to leave the oceans now, away from the hot vents and the depths of the Pacific Ocean, and take you on an adventure across the Southern Ocean to the continent of Antarctica, the coldest place on Earth. The drone of the Hercules transporter plane changes pitch. After eight hours of flying, the mind-numbing noise has sent us into a semi-stupor of sleep, but the change in engine noise wakes us. Suddenly, we feel the plane drop. Sitting in a net with little comfort and no windows, we care little about what is going on any longer, but soon the shudder of the plane and the roar of engines gives us the first indication that we've landed. Our fellow passengers look around

FIGURE 9. Deep drilling to Lake Vostok in Antarctica. Russians have drilled three and a half kilometers into the ice to reach the ice above the underground Lake Vostok, a vast sub-glacial lake. As the inset shows, at these depths microbes are still to be found. This one, just a thousandth of a millimeter long, may be related to those that inhabit the lake itself. (Image of drilling, Todd Sowers, LDEO, Columbia University, Palisades, New York. Image of microbe, John Priscu Research Group.)

at each other. There's a look of excitement mixed with trepidation and some even manage a nervous smile of glee. The main door of the plane opens and the white light blinds our eyes. Donning sunglasses we make out the landscape through the small exit. We have arrived at our destination, Vostok base, Antarctica (Figure 9). To the horizon is the white of snow; a bleak and windswept environment. Vostok is often held to be "the coldest place on Earth". At 78° S it had a temperature recorded of –89.2 °C in July 1983, the record from which it gets its dubious low-temperature accolade. Built in 1957 during the International Geophysical Year, the Russian base is home to about twenty-five Antarctic research and support staff during the summer months.

We are here because scientists have become excited about something they have found three and a half kilometers under the ice – a giant lake. The lake is a body of liquid water trapped in the ice. It is about one hundred and twenty kilometers long and fifty kilometers wide. The one million year old lake may contain microbes trapped within it when it first formed. More sensational claims have been made that perhaps microbes here have taken a completely different evolutionary course from anywhere else. Aside from the wild speculation about what may or may not be in there, there is one thing that is for sure. At these depths, as in the crust of the Earth itself, we find life. The drill cores that were first returned from a depth of three kilometers – some 300 meters above the lake – revealed microbes. The ice is, of course, frozen, so these microbes may not be very active; although it has been suggested that channels within the ice containing salt water that isn't frozen could provide a habitat for actively growing microbes, even at these depths.

How did the microbes get there? That isn't really known. Perhaps they were trapped in the ice when it first formed, perhaps they slowly percolated down into the ice as snow melted on the surface each year, perhaps a combination of these. It is known that a whole variety of microbes are blown across to Antarctica from the Americas and probably from further afield as well. At Vostok station in 1975, Russian researchers discovered bacteria and fungi, and other microbes that were blown to Antarctica by winds from lower latitudes. The numbers of the organisms at different depths in the ice, and thus different ages of the ice, change with periods of climate change on the Earth. The ice is a time capsule, preserving specimens of life as far back as one million years ago and maybe earlier.

This life, whether it is dormant or active, is protected from the uncertainties of the surface world, essentially locked into a freezer where it will remain until it leaches out in trickles of water or until the ice sheet of Antarctica melts.

So now you are beginning to get the picture. There are microbes eleven kilometers deep in the oceans, microbes three kilometers deep in the ice, microbes five and a half kilometers deep in the ground, microbes seventy kilometers up in the sky and everywhere else in between. The Earth is a ball of rock, teeming with life. The only way to make the microbial world extinct is to boil the whole planet. Otherwise, as we will now go on to see, the effects of even some of the most calamitous events that the Galaxy has to offer will barely touch the totality of the microbial world as it progresses on the galactic journey.

4 The record of catastrophe

The realization that catastrophes from the Earth or space might change the direction of life on Earth did not happen overnight. It took at least a century for the observations of rapid change in the types of fossils occurring in different rock strata to catch up with the deep cultural legacy that was imbedded in our view of the world. The notion that history was unchanging, that animals, plants and microbes were the products of a perfect process of divine design was impossible to resist in the nineteenth century and in previous centuries. The way in which an ant milks an aphid, the sonar that a bat uses in the dead of night to catch a moth, the speed and agility of a leopard in pursuit of its prey. It is not difficult to see how early naturalists and geologists alike came to the conclusion that the biosphere was quite perfect, each animal fitted into a great picture with an uncanny precision and its part in an endless cycle of predator and prey seemed to suggest that the entire system was the product of an endless tweaking and nudging to perfection. On inspection the biosphere seems to have little redundancy. Every scrap of waste is food to another animal and the entire system does not seem to be running out of control because of imbalance. How could this be anything other than a reflection of the unchanging perfection of nature over time?

The boundaries that paleontologists found in the fossil record, that suggested a rapid change at particular times in the past, were regarded for the most part as a failure of preservation, an illusion of geology and in some more fundamentalist religious circles, as a downright devious ploy by the devil to send us down the wrong track. Of the scientific camp, Charles Lyell (1797–1875) most vigorously pursued the uniformitarian view of geology, a world of unchanging certainties that he set out in his work *Principles of Geology*. To be

MORE ABOUT EXTINCTIONS IN THE FOSSIL RECORD

Sediments in lakes, rivers and oceans can be laid down at an excruciatingly slow pace. When they eventually get compressed into layers of rock that end up being studied by paleontologists, many millions of years can be represented in just a small layer, maybe just a few centimeters thick. Proving that an animal or a group of animals went extinct very quickly can be difficult to do as it might have occurred over millions of years however rapid it may appear in the fossil record. Paleontologists are therefore faced with the task of trying to find the best sections of rock they can for a particular period in Earth's history when they think an extinction occurred. They try to examine very carefully what happened either side of the extinction boundary. It is not surprising that even today people argue about whether some extinctions were gradual or sudden.

fair he wasn't the only person following this line of reasoning. James Hutton (1726–1797), a British physician, was a powerful eighteenth-century proponent of a gradualist view of the world.

Catastrophism wasn't exactly a new idea in the nineteenth century, it was just a maverick idea, the sort of thing you didn't talk too much about if you wanted to get ahead. There were figures who had the position to enable them to begin to open up a catastrophist debate. Baron George Cuvier (1769–1832), a French geologist, began to explore the ideas of catastrophe. Sea level changes, the change of the flow of rivers and the sudden changes of layers in rock beds, he observed them all. This evidence seemed to him to suggest periods of sudden change. Cuvier was also religious. He accepted many of the arguments about the formation of the Earth sometime around 40,000 BC and the possibility of a single creation, but that did not seem to affect his notion that, at least since creation, geological processes had occurred in irregular spasms of change.

The notion that evolution is not a continuous, gradual process came most eloquently in the twentieth century in the form of the theory of punctuated equilibrium, devised by Stephen Gould at

Harvard University and Niles Eldredge, now at the Natural History Museum in New York. The theory is complicated and its various nuances form the basis of much wider discussion, along with its competing theories, but its essential message is quite simple. Evolution is not a gradual process of one species making way for a new species, but can be static for a period and may then hit a period of rapid change, whatever that mechanism of rapid change might be.

It has long been recognized, and in fact even assumed, that some of these rapid changes observed in the fossil record may be wrought by biological changes, not necessarily physical catastrophes like asteroid and comet impacts (Figure 10). When a population of an animal gets very small it is more likely to be thrown to extinction by a small glitch, like a disease. Problems such as inbreeding and other genetic effects can start to manifest themselves and make the population unhealthy, so there is a minimum viable population of any organism. This can be as few as perhaps tens or hundreds of individuals, depending on the species. This is why we worry when we find ourselves, through our wanton destruction of habitats, responsible for pushing a species down to a small population size. Simply existing is not enough.

Charles Darwin (1809–1882), of course, did not know about DNA and genes. He assumed that extinction was caused by species competing each other into the ground, so to speak, rather than being caused by physical catastrophes. The dominant idea amongst the natural historians of the time was that superior species would push inferior ones out of their niches and, in so doing, extinction would occur. This idea isn't wrong, in fact some argue that plants become extinct mainly by this process, but the idea alone fails to take into account the role of physical catastrophes. It would be the catastrophists who would open our minds to the idea that sudden (geologically speaking that is) physical changes caused by say, volcanoes or asteroid and comet impact events, could bring about the end of species as well.

In the twentieth century our understanding of the cosmos and the catastrophes with which it could threaten the Earth began to

FIGURE 10. When bad things happen to good animals. The two impressive animals shown here are now extinct. Their large sizes and surface-dwelling attributes make them easy targets for various extinction mechanisms. A *Megaloceros*, a giant-antlered deer, and two woolly rhinoceros, that went extinct about 10,000 years ago, perhaps at the hands of climate change with some human help.

converge with the observations of the fossil record. It was understood that asteroid and comet impacts could happen and it was shown that rapid extinctions had occurred in the fossil record. The causative links between the mechanisms of catastrophe and extinction at any given point in the fossil record would trigger, and continue today to cause, fierce debate. However, the fact that asteroids or comets might collide with the Earth is hardly regarded as speculation any longer. The questions revolve around "when?" and "how big?", not "can it ever happen?". And the fact that the Earth has been subject to periods of catastrophic volcanism is not under much debate either. The observations of huge areas of lava that bear testament to massive eruptions, millions of times bigger than volcanoes we normally observe in our lifetimes, tell us that the glimpse of world history that we have had is not an accurate reflection of what has gone before.

A particular advocate of these types of physical catastrophes as a cause for extinction was David Raup, a paleontologist from the University of Chicago. Because the word "extinction" doesn't really make it obvious that unpleasant things might have been brought to bear on the Earth by outside agents, he preferred to talk of "killing" rather than "extinction", which is a more colorful description. He undertook a number of interesting calculations to see whether all extinctions could be explained by asteroid and comet impacts. His ideas followed on from his now (in)famous "kill curve", which predicts the waiting time between events that will finish off a certain percentage of species. Surprisingly, although Raup himself is careful to point out that the calculations on how much of the kill curve can be explained by impacts are just an interesting experiment, the statistics are quite intriguing. Based purely on the numbers, it might be easy for someone to make the case that asteroid and comet impacts have always had an important influence on extinctions (this includes some smaller extinctions, not just the big five extinctions that have happened over the last 600 million years).

Compared to the total number of species that have ever lived on the planet, which is estimated to be between about five

MORE ON EXTINCTION

The word extinction is, in itself, quite controversial. How do you define an "extinction"? For example, extinction might simply be the number of families that go extinct in a particular part of the fossil record. This is simple, but it doesn't take into account the number of families at risk from extinction. If the numbers that go extinct are just a small fraction of a massive number of species there to begin with, then it might not be very important. So you could take the number of families that went extinct and divide it by the number of families that were alive to begin with and use that as a measure of extinction and so on. Whatever method of measurement you use, the five large extinctions of the last 600 million years still pass the test and emerge as the greatest periods of biological change on the planet.

and fifty billion since life began, the fossils, which cover about 250,000 species, are a meagre percentage. However, they give us a good idea of when periods of change occurred on the planet. Over the past 600 million years there have been at least five great episodes of mass extinctions and, no less interesting, in amongst these are several episodes of smaller-scale extinctions.

I am a microbiologist, not a paleontologist, and so I am going to avoid (if only for the sake of avoiding a catastrophic debate about this chapter) lots of discussion about each of the extinction boundaries that have been found in the rock record and what may and may not have caused them. In later chapters I'll describe some of the biological catastrophes that might be caused by asteroids, comets, volcanoes and even human activity. Although I will talk about the link between some extinctions and these catastrophes, such as the supposed link between the end of the dinosaurs and an asteroid impact 65 million years ago, I will leave this debate to other places.*

---

* A good place to read about this debate is in the book *Controversy: Catastrophism and Evolution: The Ongoing Debate* by Trevor Palmer published 1997, Kluwer Academic/Plenum Publishers.

My purpose here is to provide context. I want to give you a flavor of what is observed in the fossils for a journey once around the Milky Way Galaxy. We will start 250 million years ago, when Earth was approaching the same position in the Galaxy that we are in now. We will take a journey once around the Galaxy lasting 250 million years and see what happened to life on Earth as is revealed by the fossils. As the book continues we can begin to ponder whether the catastrophes we experience, asteroids, comets, supernovas and volcanoes, might be the cause of some of the great changes I am about to describe. We'll perhaps begin to understand the amazing way in which life changes during the galactic journey and the amazing robustness of the microbial world.

Two hundred and fifty million years ago, the biosphere was emerging from one of the greatest crises it had ever faced. Within the space of perhaps less than two million years between three-quarters and over 95 percent of all species on Earth went extinct. The end-Permian extinction was one of the most complete that has been recorded. Some of the best records of what went on during this time on Earth come from south China. Probably about a quarter of a million species that at the time occupied the Earth were reduced to just 10,000. Even insects, that had been quite resilient to previous mass extinctions, faced a large-scale wipeout, testament to the severity of the cataclysm. After the end-Permian the way would be cleared for completely new types of insects to emerge, like the beetles. Marine organisms, like the corals, suffered badly as the reefs of the world underwent a complete collapse, one of the most complete of extinction events. They disappeared from the fossil record for about eight million years.

Of the microbes in the sea, the tropical varieties were affected the worse, but ones that fed off rotting plants and animals and other remains of other organisms seem to have survived quite well. A spectacular group of microbes, called radiolarians, suffered extraordinary losses. These microbes, with their glassy ornate skeletons, are an ancient group dating back to 600 million years ago. Despite their long and illustrious history, they suffered an unmitigated disaster. Microbes with soft organic walls, the acritarchs, that prefer poorly oxygenated waters that other organisms would find stressful, seem to have got through quite well.

It is unlikely that microbes in the deep subsurface of the Earth were affected at all. We have no fossil record of subsurface microbes from this time, but for microbes such as those that use hydrogen and carbon dioxide to live, events on the surface would have been irrelevant.

Some species of animals make very good indicators of extinctions. They are quite fussy about their physical requirements. They might like particular temperatures, for example. However, they are abundant, and so if they die they leave a very good trace in the fossils. Ammonites, which are marine molluscs some of which have spiral shells, are a very good indicator of change and they suffered an almost complete extinction at the end-Permian. One hundred and two species out of 103 were wiped out. Trilobites, hard-shelled segmented

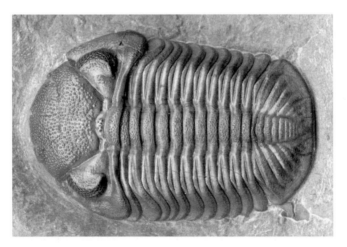

FIGURE 11. Trilobites went extinct at the end-Permian about 250 million years ago in the greatest extinction recorded, when between 75 and 96% of all species became extinct. These arthropods, which ranged from one millimeter to one meter long, held a variety of niches, including planktonic feeders and swimming forms. They probably numbered about 15,000 species and existed for over 200 million years, one of the most successful organisms recorded in Earth history. (Image, Sam Gon III.)

creatures (Figure 11), that had dominated the Earth were made completely extinct. It is very clear that something profound was going on in the oceans. There is one mystery about the oceans; the sharks and fish seem to come through almost unscathed. Maybe they fed on the decaying remains of the other organisms that were completely destroyed. Sharks are a primitive group and although they are not very intelligent, they can eat anything. Perhaps their primitiveness is testament to their resilience to environmental changes including those that hit the Earth 250 million years ago.

The land didn't go unscathed either. Plant fossils suggest an unusual crisis occurred. Ferns and weeds replaced many of the large plants of the time, like the peat-forming trees in China. Weeds like the quillwort, *Isoetes*, made it through and did well after the extinction. Most fascinatingly, the fossil record reveals that during the end-Permian extinction the number of fungal spores sky-rocketed. The planet became covered in fungi, like a rotting apple in an orchard. Is this just a symptom of a global mass die-off?

Over half of all the large surface-dwelling animal families went extinct as well. The large plant-eaters and six out of the nine amphibian families were gone forever. The reasons for their demise are not so clear. We do know from the chemical record that photosynthesis in the oceans was much reduced during the time of the extinctions and land photosynthesis might have suffered as well. Did a global shut down of photosynthesis shut down the supply of energy to the bottoms of the world's food chains and cut down the food available to large animals?

The end-Permian wasn't just one big mass die-off, in fact it came in at least two episodes, but what could have caused such a vast extinction of animals and plants on land and in the sea? The reasons for this huge collapse of the Earth's biosphere, the greatest of the last 600 million years, are still being debated. One favored hypothesis is that the atmospheric levels of the greenhouse gas, carbon dioxide, increased and caused a period of global warming. The gas may have come from volcanic eruptions in what is now Siberia. These "Siberian

Traps", of which we will hear more later, would have ejected millions of cubic kilometers of ash and lava, contributing to the build-up of this greenhouse gas. The global warming event might have had two consequences. It would have reduced the amount of oxygen that could dissolve in water (because less oxygen can dissolve in water as the temperature increases) and it would have slowed down the transport of water from the equator to the poles. And so the argument goes, the oceans simply went stagnant and got depleted in oxygen. The idea is supported by observations of the rock types laid down during the extinctions. A symptom of oxygen-poor conditions are black shales that are found to overlay white clays. And there is a lack of cesium depletion. In oxygen-rich waters the element cesium is normally scavenged and bound up by iron oxides in the water. If there is little oxygen then there are much lower concentrations of oxides and so the cesium doesn't get mopped up. Higher concentrations than normal can suggest that the oceans lost their oxygen.

Like all extinctions, there is no end of possible hypotheses to explain why so many species might have gone extinct. Some propose that the volcanic winter produced by the ejection of dust and ash by the Siberian Traps could have cooled the Earth, darkened the skies and shut down global photosynthesis. The loss of more tropical species than polar ones, that would be better able to survive cold, is taken as evidence that the Earth cooled during this time. Although it is possible that the Siberian Traps might have caused volcanic winters during particularly heavy eruptions, the end-Permian seems to be associated with global warming. The chemical evidence suggests that temperatures could have increased by as much as 6 °C. Some cold-adapted plants seem to be replaced by ones that prefer warmer conditions.

Then of course there is the impact scenario. The physical changes at the end-Permian don't entirely fit the idea of an asteroid or comet impact being responsible, but some researchers have found a rapid change of sulfur and strontium chemistry at the end of the Permian, which they think was caused by the collision of a massive object with the Earth. It seems to come along with a high concentration of shocked rock grains that might be formed during the impact. For the end-Permian the impact scenario is a newcomer on the scene, but its involvement cannot be ruled out and it might have just been one prong in a multi-pronged attack necessary to ensure the demise of over three-quarters of the world's species.

These aren't the only theories. Others contribute a variety of ideas, such as the oceans becoming salty thus killing everything that couldn't tolerate the salt; a supernova explosion killing everything from cosmic radiation (although there is no chemical evidence for this); a global pandemic of an unknown disease; or perhaps some unknown chemical poisoning the oceans.

The biosphere would recover from this traumatic event, but fifty million years later, just a fifth of the way further in the Earth's journey around the Galaxy, the biosphere would be struck by yet another extinction event – the end-Triassic extinction. Less extreme than the end-Permian, this event may have been just as dramatic in terms of time scale. Over a period of four million years, and possibly as short as one to two million years, just under half of the invertebrates would go extinct. The microbes of the sea suffered, but with a great variation from group to group. Some, such as the radiolarians, the small ornate organisms that we saw suffered devastation at the end-Permian, didn't seem to be affected, but others, like the swimming dinoflagellates, did suffer extinction. The large surface-dwelling animals seem to have escaped relatively unscathed, but many plants, particularly some families of ferns, seem to have gone extinct. Some researchers have presented evidence that the extinction saw the emergence of some of the first dinosaur families that would later rise to global dominance and the demise of various amphibian and mammal species, but not everyone agrees with this.

As with any other extinction, the causes of the end-Triassic extinction are a subject of contention. The volcanic camp has claimed a link between the extinctions and the eruptions of the Karoo volcanic region in South Africa, but it is possible these eruptions were too late for the extinctions. The claim that the Manicouagan impact crater in

Canada is testament to the arrival of an asteroid or comet at the time seems to have the problem that the date of the impact crater at 214 million years is a little bit too old, although that doesn't rule out other impact events being involved in the extinction. A favored hypothesis is that there was a large-scale drop in sea level followed by an increase in sea level with oxygen-poor waters, causing a destruction of habitat in coastal regions around the world with its ensuing effects on life. Support for this idea is found in the form of rocks that suggest shallow waters turning rapidly into oxygen-depleted waters, manifested as black shales on top of rich limestones. Along with the change in the rock type is also an abrupt change of the animals inhabiting these regions. Many animals that lived in shallow-water reefs met their end, including a whole variety of sponges.

Often the fossil record is scanty and millions of years may be represented by just a few centimeters of fossil layers. Was extinction sudden or very prolonged? When you are trying to find out from a layer with a thickness of a few centimeters or less, it is very difficult to know. What might look abrupt might be an effect occurring over hundreds of thousands of years which, as you'll appreciate, is abrupt in geological terms, but is not as abrupt as, say, an asteroid impact, which would be expected to cause global devastation in a matter of days to weeks.

There is a human dimension to trying to find out what killed off past animals as well. Good scientists don't allow themselves to be affected by personal prejudices and seek only the truth. In reality of course things aren't this simple. After carrying out many years of research on particular mechanisms of extinction, it can be very difficult to move to new research. To invoke volcanism as a cause for the end-Triassic extinction, you need to know all about volcanoes. You need to know about their physical, chemical and biological effects. You need to know where in the world they occurred. You need to have studied many volcanic samples to be able to know what geological evidence you are looking for and of course, more pragmatically, you need funding to get to your field site. Now if you decide to agree that asteroid and comet impacts might have caused the extinction, suddenly you are in a completely different ball game. You need to know about asteroids and comets. You need to know how frequently they occur and what geological evidence they leave in the fossil record. You need a laboratory that has the expertise to answer these questions. And so you can see how different "camps" get set up; the "volcanists" and the "impactists". Despite this cynicism, generally it seems to work quite well and different arguments do get heard. The uncertainty of what has caused different extinctions just shows how far we have to go in studies of extinction and their mechanisms.

No mass extinction boundary has captured the imagination as much as the one that next hit the biosphere. When the Earth was three-quarters of the way around the Galaxy and just 65 million years short of our current location, this small dot would suffer yet another crisis. The Cretaceous–Tertiary extinction (K/T extinction) was not as great as the end-Permian that I described earlier, but it was associated with the demise of the dinosaurs and so it captured not just scientific interest, but the imagination of the public. Never have so many books been written about a single extinction boundary and never has such a one-sided view of an extinction and its mechanisms been presented to the public before.

I don't want to just focus on dinosaurs and extinction, which is why I have described some of the other extinctions that have happened over the last 250 million years. The end-Permian is just as interesting as the K/T – and bigger! My purpose is to set the historical stage before we look at the survival of the microbial world during catastrophes.

The K/T boundary is certainly interesting, not so much because of the demise of the dinosaurs, that up until then had held supremacy of the air, land and sea for 120 million years, but because the boundary has been fêted as having the greatest evidence associated with the impact of an asteroid or comet (Plate V).

Sixty-five million years ago more than three-quarters of all species disappeared. Of all the species that went extinct at the time, dinosaurs are the best known, but actually the most problematic.

Their bones are large and not very common, so it is difficult to tell whether they went extinct suddenly by looking at the changes in fossils. Some have said that the dinosaur extinctions were already happening by the end of the Cretaceous and that, in fact, they were beginning to go extinct at least seven million years before the boundary. Their demise might have just accelerated during the last 300,000 years. Others have said that the dinosaurs had very high turnover of species anyway and that species were constantly going extinct and being replaced by new ones. What happened at the end of the Cretaceous is that there were no new ones to replace the old ones and that brought about their demise. Perhaps biological factors were as much to blame as any physical factors.

The idea that the end of the Cretaceous was heralded by some poor dinosaurs on a plain somewhere looking up at an asteroid flashing by and roaring as the sky lights up with the blinding flash of the instantaneous calamity that will destroy them has become deeply rooted in public perceptions and Hollywood films. It is a chilling, somehow fascinating idea. The extinction of the dinosaurs may have been much more complex than this though.

What is so interesting about the K/T extinctions is that the fossil record is a lot better than for the earlier extinctions, so we can begin to see what selectivity the extinction mechanism may have had. Most immense dinosaurs and tiny mammals were equally affected by the events. About half of the mammal species did manage to survive.

It does seem as if the animals in the oceans fared worse than freshwater species that lived in lakes and rivers. Crocodiles and amphibians made it through, but the marine dinosaurs vanished. Animals living in the water as opposed to those that hide out in the silt and sand of the continental shelves seemed to have been much more badly affected. An exception to this appears to be at the high latitudes and polar regions. Some scientists have vigorously jumped on this fact to suggest that indeed the Earth was smothered in an impact winter, when dust and soot generated by the asteroid collision circled the Earth and cut out the light. Organisms at high latitudes that were tolerant of cold conditions and polar darkness would have survived these events.

Like the end-Permian, plants were also hit. The flowering plants were worse affected, by contrast conifers made it through. Incredibly, the abundance of fern spores goes from about a quarter of all the spores in the fossils before the extinction to close to a hundred percent just after, suggesting that the opportunistic ferns dominated many areas of the world, just as they do in disturbed or felled forests today.

What grabs the attention of the paleontologist about this particular extinction is the geological evidence that goes along with the biological changes. Shocked quartz, pieces of tiny quartz granules that have the diagnostic features of being subjected to a sudden and very large shock pressure, suggest that an asteroid or comet hit the Earth at this time. The now-famous paper in the journal *Science* by Luis Alvarez in 1980 showing that the chemical element iridium has a higher concentration at the boundary was taken as evidence for an impact because an extraterrestrial object could deliver this element to the surface of the Earth.

And then there is the smoking gun, as it has so often been described. The 180-kilometer diameter Chicxulub (pronounced Chickshoo-loob) impact crater in the Yucatan Peninsula, Mexico. It dates almost exactly to the boundary and it is the size that would be expected from the 10-kilometer diameter object needed to cause the extinctions. A deposit three meters thick in north-east Mexico has been claimed to be the remnants of a giant tsunami deposit thrown out across the oceans by the colliding object. The mixture of deepwater sediments can be dated almost exactly to the K/T boundary and it isn't the only tsunami deposit to be found. And so despite the counter-arguments, it must be agreed that there is a great deal of evidence for an asteroid impact contributing towards the demise of the dinosaurs and so many other organisms at this period in Earth history.

How would such an asteroid have killed so many animals, plants and microbes? When the asteroid hit the Earth it would have injected dust and soot into the atmosphere, shutting out sunlight around the globe, maybe for six months. Possibly, if large amounts of sulfur had been injected into the sky, for years. Some say that the demise of the polar dinosaurs that were well adapted to cold and dark shows that the impact winter hypothesis cannot be correct. However, most polar animals of today depend upon the short summer season to stock up on food. If the impact winter had adversely affected even one polar summer season and prevented the short growing season from happening, it is possible that even cold- and dark-adapted animals would have suffered the knock-on effects. The impact scenario doesn't end with the winter.

The decline of the Foraminifera at the boundary, small microorganisms with ornate shells made from, among other things, calcium carbonate, has been suggested as evidence of acid rain. Acid rain is formed in the atmosphere by the heating of the atmosphere by the incoming asteroid. The acid, which would have rained into the oceans and acidified the water, would explain why the acritarchs, which have organic walls and are less susceptible to acid, seemed to survive relatively well. The idea is problematic because freshwater animals, like the amphibians, are very susceptible to acidification of the water and yet they survived. And there are some other mysteries as well. Birds seem to have survived quite well. One would expect birds to have been badly affected by the calamitous effects of a massive asteroid and comet impact.

Although the evidence for an impact event is quite good, could other factors have been responsible too? Was the impact event the *coup de grâce* in an already stressed biosphere? Like the Siberian Traps at the end-Permian, the Deccan Traps of India seem to coincide with the extinctions. Did the massive eruption of basalts from deep within the Earth cause the extinctions? The Traps seem to have been erupting at least a few million years before the extinctions, so it is possible that they were piling stress on the biosphere, perhaps by increasing atmospheric levels of carbon dioxide and causing periods of warming. This may have been interspersed with very short periods of cooling caused by light-blocking dust and ash ejected by the volcanoes.

Alongside these stresses there may also have been changes in sea level. Rocks found in Denmark suggest that sea level was fluctuating up and down before the extinctions and for some time afterwards. Dropping sea levels would cut down the flow of nutrients into the shallow waters of the oceans and cause a crisis in the animals living there. Shallower seas might have increased the number of freshwater inland habitats, like lakes and ponds, which would have improved the chances of survival of freshwater animals, like crocodiles, exactly what is observed.

Sea level changes also cause a more interesting effect to kick-in, called the species–area effect. On a given plot of land only a certain number of species can be sustained. If you took a small plot of land that had a whole zoo of animals feeding on it and put a fence through the middle, both plots of land, now being smaller, can sustain fewer species and so some of the species on both sides of the fence would become extinct. Biologists have been studying the species–area effect for a long time on small islands and it is one of the most fascinating ways in which the physical environment can exert an influence on extinction. If the sea level goes down, then two things happen. The area of shallow waters and continental shelves goes down and that means fewer species can be sustained. There must be extinctions. However, the land area increases because it becomes exposed by the receding water and so the number of species that can live on land can potentially be increased and, of course, vice versa. So periods of extinction could be triggered by sea level changes simply by the effects of changing land area, quite independently of the direct effects of the sea level change itself.

And so, whilst trying to avoid the error of becoming carried away with one extinction mechanism or another, it should be apparent that there is great debate in these matters. I have tried here to present some of the ideas that are in the literature today. On balance the evidence for an impact event looks good, but it also seems as if it might have been just one stress in a biosphere already facing changes at the time. Changes in sea level and volcanoes may have set the biosphere up for a crisis once an asteroid impact did occur – and this is just one interpretation.

That extinctions do occur is undoubted and the evidence for some of these being caused or accelerated by catastrophic events, volcanoes and impact events in particular, is getting better all the time. The microbial world seems quite robust against these changes, but not immune. The devastation of the Foraminifera at the end of the Cretaceous and the destruction of the microbial communities in the corals of the end-Permian are just two examples of microbial communities that went the way of the large, extinction-prone, surface-dwelling animals. However, these are invariably the microbial communities that depend upon photosynthesis. The capture of light as an energy source brings with it the risk of extinction because of the need to live near the surface. At the end of the Cretaceous, there is no doubt that five kilometers under the ground, the hydrogen-using microbes of the deep subsurface continued their lives uninterrupted by the calamities unfolding at the surface. Their family histories have completed one circuit around the Galaxy oblivious to the upheavals that have been occurring on the surface.

Although we can say, almost for certain, that groups of microbes will make it through the most extreme extinctions, we can probably never know with certainty exactly which species will do well and which ones are likely to go. We still know so little about even present-day ecosystems. How would a period of acid rain 250 million years ago affect a polar lake compared to a tropical one? How would a volcanic winter 65 million years ago affect tropical and temperate forests? The study of the effects of climate change on present-day ecosystems is a vast area of research, driven along by our interest in predicting the possible effects of mankind on the biosphere. Slowly, but surely, we chip away at the corners. We find out what effect a rise in carbon dioxide will have on a particular species of rainforest plant. We investigate how a doubling of ultraviolet radiation caused by the ozone hole would affect a strain of rice plant and we discover how a change in temperature might affect a plot of land with a few different species. Turning this into an understanding of how the present biosphere changes, how the millions of interactions betweens plants, animals and microbes are affected by a change in even a single environmental factor is a vast undertaking.

Perhaps we will never be able to fully predict how a continuous environmental change alters the biosphere *in toto*. Perhaps we can only know by observing it for ourselves as it happens – even more of a reason to mitigate our potential influence on the environment. And it is for this reason that our chances of dissecting and understanding properly how past environmental changes would have affected a biosphere populated by organisms that are now long extinct are virtually nil. You cannot experiment on extinct animals. The best we can do is grapple with general concepts and understand the general direction that the biosphere might go in. Darkness will shut down photosynthesis. Sea level change will kill certain groups of marine microbes. Perhaps we can begin to explain how broad groups of organisms were affected by large-scale environmental changes, but a true understanding of how and why particular species went extinct and others did not may forever be covered by a fog of incomplete data. Ironically the extinction of species itself may deny the scientific community an understanding of extinction.

5   The sky falls in

Even as recently as the mid-twentieth century, many regarded the notion that rocks could fall from the sky as preposterous. The absurdity of the idea was based partly on the fact that there was little empirical evidence that stones had ever fallen from the sky, and in view of the fact that scientists need evidence before coming up with theories, this was quite reasonable. The less reasonable block on the idea came from the fact that up until the mid-1800s when Darwin, Wallace and others would change our view of the origins and *modus operandi* of the natural world with the theory of evolution, it was preferred that all theories fitted the ideas laid out within the Bible, word for word. So the fact that God was the creator of heaven and Earth and the Earth was formed sometime around 40,000 BC, or maybe even exactly in 4004 BC, didn't fit too well with any theory about leftover remnants of an ancient Solar System episodically falling in to Earth and destroying large percentages of God's perfect, unchanging creations. With the underpinnings of this religious legacy, the uniformitarians held sway.

Proof of the biological calamity that could be wrought by the leftovers of the imperfect process of planetary formation should go, in honor and respect, to the surprised citizens of the small Russian region of Tunguska in Siberia, who at 7.14 AM on June 30, 1908, were personally privy to the destructive effects of an asteroid or comet on a collision course with Earth. For on that day a stony asteroid or comet exploded in the air near their village. In an instant, an explosion one hundred times more powerful than the nuclear bombs detonated in World War Two lit up the sky and burned and flattened over one thousand square kilometers of forest. The fine dust wafted up into the atmosphere by the explosion reflected light around the world. Across Europe newspapers could be read at midnight.

Little was done to find out what had flattened the forests until 1927, when the first Russian expedition visited the site and found the remains of the burned trees. Laboratory experiments suggested that an airburst explosion from an extraterrestrial object could have flattened the trees as the experiments, using charges exploded over thousands of upright matches stuck to the ground, exactly emulated the butterfly shape of the destroyed trees. In the 1990s, Italian–Russian expeditions to the site re-invigorated interest in Tunguska. Using seismic data collected by stations across Europe at the time of the explosion, coupled with new studies of the explosion site, they believe that the object came to Tunguska from the south-east at about 11 kilometers a second. At about 10 kilometers above the ground it broke up, perhaps because it was an asteroid made up of loosely bound together fragments of rock that disintegrated before it reached the ground.

Fortunately for the residents of this small village, they were far enough away from the event to be largely unscathed. There are stories of some nomads who may have been in the blast zone at the time of the explosion and undoubtedly some animals succumbed to the destruction at ground zero, directly under the explosion, but the biological casualties were minimal. If the object had been just a few hundreds of meters wider, it would have completed the final 10 kilometers to the surface of the Earth and a crater would have been carved from the ground. The blast wave would have swept the village and the residents away. And they weren't the only lucky ones. If it had happened only six hours earlier the object would have exploded over St Petersburg and destroyed the city.

David Kring, of the University of Arizona, did a nice study in 1997 of the effects of the object that slammed into the deserts of Arizona 50,000 years ago forming the spectacular one-kilometer diameter bowl-shaped Meteor crater just east of Flagstaff. With an energy perhaps a thousand times greater than that at Tunguska, the impact created 2000 kph (1240 miles per hour) winds in the blast wave that were so intense that hurricane-force winds would still have been felt 20 to 40 kilometers away from the crater. Trees and animals would have been swept from the ground and carried through the air for many kilometers. Fortunately for us, Meteor crater-sized impacts happen only once every 10,000 years, that's 25,000 times during our journey once around the Galaxy. During the history of our species, *Homo sapiens*, such events have probably happened about 200 times.

The Pretoria Salt Pan, or Tswaing as it is more often called now, is another impact crater about one kilometer in diameter. Just north-west of Pretoria in South Africa, the imposing hole in the ground has a lake in the middle and the lush greenery of the African bushveld surrounds it. The crater was formed about 200,000 years ago, when primitive humans were still wandering the plains of Africa. The blast wave would have done much the same as that calculated for Meteor crater. Imagine what our ancestors must have thought when one peaceful day in Africa the sky erupted with the explosion of an asteroid impact possessing the power of the present-day world's nuclear arsenal detonated all at once. These impacts, powerful and frequent enough to have had devastating effects on life near to them, but small enough not to have threatened the existence of humanity, must have had profound effects on early human civilizations and their views of the world.

As decades progressed and astronomers examined asteroids and comets hurtling through our Solar System, scientists would begin to realize that Tunguska-size events happen about once every 100–1000 years. In other words, there are at least a quarter of a million Tunguska-sized explosions during our journey once around the Galaxy. However, their high frequency is matched by their low destructive power, at least in the global context.

It would be the geologists, led primarily by Luis Alvarez, who would add another, more extreme, data point to our perception of what asteroids and comets can do to life on Earth, by publishing a paper in the journal *Science* in 1980. As discussed in the previous chapter this paper presented geological evidence that the demise of the dinosaurs and over three-quarters of life on Earth at the Cretaceous–Tertiary (K/T) boundary 65 million years ago was caused by an asteroid.

The most fêted evidence for this impact, and the basis of their *Science* paper, was the high concentrations of the rare element iridium, found at the K/T boundary. Its concentrations were about 5 parts per billion, over a hundred times higher than it would normally be in the Earth's crust, suggesting that something had delivered large quantities of iridium to the Earth, in fact about a quarter of a million tonnes of the element. Some meteorites have high concentrations of iridium and this seemed a likely source of material. Based on the concentrations of iridium in meteorites, one can work out the size of object that might have caused the extinctions and it was found to be about 10 kilometers in diameter.

Meteorites aren't the only source of iridium, though. Iridium was later found being emitted by the Kilauea volcano in Hawaii and some scientists began to suggest that rock from deep within the Earth, erupting at the surface, might be able to deliver high concentrations of iridium to the surface. Perhaps giant volcanoes could throw iridium into the atmosphere. As we saw in the last chapter, the controversy of volcanism versus an asteroid or comet impact for the demise of life at the K/T boundary continues. There is other evidence that supports the idea of an impact being involved. Quartz with features of high-pressure shock and small glassy spheres or "tektites" have been found at the K/T boundary in most rock sections around the world. They suggest that something created massive pressures and ejected vast quantities of molten rock, distributing it around the world. These observations are consistent with the impact hypothesis.

With the weapon now apparently identified, the site of the event became the next focus of the scientific quest. From the amount of iridium in the layer and the conclusion that the object was about 10 kilometers in diameter, the scientific community knew that it was looking for a crater about 200 kilometers in diameter. It is not easy to hide something of this scale. In the early 1980s petroleum geologists working on the western edge of the Yucatan peninsula in Mexico had found a 200-kilometer diameter crater during gravitational studies of the area. Being geologists focusing on trying to find oil and valuable minerals, they made only passing reference to the demise of the dinosaurs and it took another decade for scientists to hypothe-size that this crater was linked to the K/T extinctions. Further stud-ies of this, the Chicxulub crater, by drilling showed that it was about 65 million years old. The age matched the K/T extinction. If this really was the impact that wiped out the dinosaurs it happened in shallow water on the edge of continental crust.

Scientists can begin to get an idea of how often impacts on this scale happen by examining the size of craters and their abundance on the surface of the Moon and even on Mars. The Moon has had very little alteration on its surface over the lifetime of the Solar System because it has no atmosphere to erode rocks away and the surface has no active volcanism or plate movements. Mars has been more eroded by wind, and even water in its early history, but it still has a very good crater record. The craters provide a record of how of-ten the planets in the inner Solar System are pummelled by aster-oids and comets of various sizes. An asteroid about 10 kilometers across that is capable of causing extinction on a par with the evidence at the K/T boundary has an energy of about ten billion Hiroshima nuclear bombs and has a frequency of about once every 100 million years.

The chances of you living to see such an event are about one in a million; the chance of a typical species being involved in such an event is about one in ten. But, during the journey around the Galaxy, the Earth will be hit roughly twice by such an asteroid or comet – one of a size that is hypothesized to be able to cause mass extinctions. It would seem that extinction-causing impact events are a fact of life during the galactic journey.

You will understand that statistics is just statistics. An event with a statistical frequency of once every one hundred million years could happen today and it could also happen tomorrow and then not happen again for another two hundred million years. Statistics tell you the probability that something can happen, they tell you nothing about whether it actually will. Although the cratering record told us something about the frequency, a question then arose as to whether there was a periodicity in impacts. Were impact events like buses; were there none for ages and then they all turned up at once? Because of our galactic journey, which takes us in and out of the galactic plane, it might be reasonable to hypothesize that there could be periods when the comet-filled Oort cloud would be perturbed and send life-threatening comets into the Solar System and then other periods of relative quiescence that might be driven by the variability in our journey.

During the 1970s this became a popular idea, but there was little knowledge of the frequency of impact events from which to try and extract evidence. By examining craters on the Earth a number of teams suggested that there is periodicity, claiming that impacts happen with particular regularity about every 30 million years. To explain this observation planetary scientists once suggested that perhaps the Sun was a double star. Every 30 million years or so our dark and distant companion, named "Nemesis", would perturb the Oort cloud, sending comets into the inner Solar System. Further investigations of cratering rates on the terrestrial planets don't seem to suggest any periodicity and today the idea, although still interesting, is not really supported by good evidence. No astronomers have found evidence for a distant companion to the Sun and there is no good support for the fact that the frequency of impact events is linked directly to our position in the galactic journey, although this is still an open and hotly debated topic. The fact that our Solar System will experience periods when it is more influenced by the gravitational forces of near-by stars as we move through the Galaxy is undoubted and so it would not be surprising if these encounters did in fact cause perturbations in our comet cloud.

Uncertainty about exactly when such events happen has driven many scientists, and recently politicians, to set up programs to try and observe the numbers and types of objects in the Solar System that cross the Earth's path. This approach is more direct than just looking at the craters these objects have left behind in the past. Using a number of

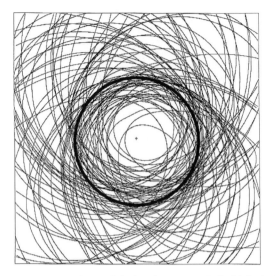

FIGURE 12. The Earth (orbit shown as a dark thick circle) is like an apple surrounded by angry wasps. Each track around the Earth is one of the hundred largest Near-Earth Asteroids that orbit our planet. (Image, Richard Binzel.)

telescopes both in the northern and southern hemisphere, a picture is now beginning to emerge of the astronomical environment in which we humans live. So far we don't know of any asteroids on a direct collision course with Earth, but as you can see in Figure 12, we are surrounded like an apple surrounded by angry wasps.

The purpose of efforts to track these objects is to give us warning about when they might happen, so society can either prepare itself or try and divert the oncoming threat, perhaps with missiles. The true efficacy of these diversion schemes is still rather controversial.

The study of the K/T boundary by a number of teams around the world and intensive computer-modeling efforts have begun to reveal what might actually happen if a large asteroid or comet did collide with the Earth. Luckily for us, the effects of large impacts remain quite speculative and we can truly say that this is one branch of science where a lack of data and being relegated to speculation is a very good thing!

The effects of an asteroid or comet impact, of course, as we saw for Tunguska, are really very dependent on the size and speed of the object. What concerns us in this book are events that cause extinctions on a global scale, because anything that happens locally will be survived by unscathed organisms somewhere on the Earth and certainly by many microbes.

Because of the uncertainty about the K/T boundary and whether it was really caused by an asteroid or comet impact or perhaps even large volcanoes, which might produce some of the same environmental disasters, I won't start to discuss the particulars of species that went extinct and try to tie them into the effects of an asteroid or comet or I will find myself on thin ice. However, I will, every now and then, talk about some of the work that has been done by groups that claim to have found links between an impact and the environmental conditions at the time of the K/T boundary.

Collision of a rock with the Earth is a very energetic event and as well as carving out a great crater, the energy will expel vast quantities of heated material. Energy is proportional to the square of the speed, so you can understand that if you are traveling at 20 kilometers a second (that's forty-three thousand miles an hour), a typical speed for a comet, then there is a lot of energy. Crash a 10-kilometer piece of rock into the ground at forty-three thousand miles an hour and a great deal of material gets thrown into the sky. Try to imagine it.

Some of this comes in the form of molten rocks just a few centimeters in diameter or less and some of it comes in the form of fine dust. The rock and dust is thrown so high into the sky that some of it travels back into space across continental distances to re-enter the atmosphere and land back on the Earth in a remote region. By a process called "ballistic skidding", the pieces of rock can be transported globally. The problem with the hot dust and rocks is that, as well as radiating heat downwards to the Earth, they also land in forests and scrub, igniting fires. An event on the scale of the asteroid that is supposed to have killed the dinosaurs would ignite global wildfires – a global conflagration might result.

A thin soot layer at the K/T boundary found by Wendy Wolbach in 1988 attests to the possibility that these fires happened at the K/T boundary. Some regions will ignite more readily than others. Ecologists have studied the frequency of fires in many regions around the world that are started by dry hot spells or lightning. They have some idea about how flammable different ecosystems are. Dry prairies have a fire frequency of about once every year and the Alaskan pine forests about once every 130 years. The dry grass and branches of prairies and forests ignite quickly and once a small fire begins, areas many tens of square kilometers can soon catch light. During an impact event, just one red-hot fragment of rock launched into a scrubland may be sufficient to doom thousands of square kilometers to a burning blaze.

Normally when fires start animals can escape the flames by running. A few unfortunate individuals will be caught, particularly if the flame front is fast. This is the reason why you can often see hundreds of birds circling around flame fronts in forest fires – they are catching the rodents escaping the fire. An advantage of the moving flame front is that the plant life exposed to the fire is heated for only a short time; once the flames have passed, the smouldering land can begin to cool. Because it takes time for soil to heat up, if the flame front is fast enough, just a few centimeters under the ground seeds and bulbs may survive intact to recolonize the burnt ground. In amongst them a diversity of microbes will also have survived and can begin to recolonize the sterilized soil.

Impact fires could be very different. A large number of heated rock fragments could ignite forest and scrub in many places at once. Thousands of flame fronts would be simultaneously ignited and there is no obvious direction of travel to escape. The fire, rather than being just an intense front of flame, would become a great conflagration of flames, perhaps forming a fire whirlwind, such as those observed in incendiary-bombed cities during World War Two. These are fires so intense that they suck in the air around them, causing asphyxiation to animals in close proximity and generating intense temperatures, in some cases above a thousand degrees centigrade. Anything within such a fire would have no escape and the soil would cook to some depth, perhaps killing off all plant seeds and roots in the vicinity. As we have seen though, even these intense fires would be survived by the microbial world. Microbes, penetrating the Earth down to a kilometer, and in some cases down to five kilometers, would have immunity from these fires.

Fires not only directly kill animals, but anything that does survive and depends on living plants for food would rapidly go hungry. It is no good escaping from a fire, only to come back and find there is not an inch of habitat left to recolonize. So large surface-dwelling animals are vulnerable to fires, both directly and indirectly. Even in regions with no significant combustible material, the heat radiated from the globally dispersed rocks may be enough to barbecue any surface-dwelling animals. Perhaps only animals and plants underground could escape the period of impact heating to emerge.

Fires produce toxins. Noxious and poisonous compounds such as cyanides are given off in intense fires, the composition of the noxious brew depends on what is burning. As so many different plants would burn in impact fires, one can expect a witches' concoction of toxic compounds of every imaginable variety to have been released. For anything on the surface that breathed free air (as opposed to, for example, a deep-sea fish that breathes oxygen dissolved in the water), these toxic fumes would spell instant death. Of course, it isn't really known how global the distribution of these toxic substances would be. Some places, like present-day Antarctica and the polar deserts of the High Arctic will not catch light at all because there is so little vegetation. We can't just assume that air currents and wind would distribute toxins evenly across the Earth. Maybe there would be places remaining unscathed by both fires and toxic chemicals.

The regions of the Earth almost certainly remaining unaffected at this stage of the post-impact scenario are the deep subsurface and the deep oceans. In both of these places, of course, fires cannot penetrate and toxic compounds take a very long time to percolate.

A deep-sea fish, at one kilometer depth, would swim around absolutely oblivious to the calamity that has just ensued on surface of the landmasses. However, ultimately the knock-on effects of death in the shallow, productive upper regions of the oceans might affect even them. But there is a much greater significance to these regions. As we saw in Chapter 3, these are the regions that are dominated by microbes. Regions so deep that, apart from the worms and clams that live around the hydrothermal vents, they are essentially devoid of complex animal life compared to the exposed surface of the Earth. Many of these microbes, particularly those at kilometer and greater depths in the crust are oblivious to the calamity that has ensued.

Soot from the fire rises high into the sky and mixes with the dust thrown up from the impact as well as sulfates and other chemicals produced from the rocks that the asteroid landed in. This challenges life with the next menace in this unfolding environmental disaster. The sky becomes dark and light levels drop so low that plants and algae that depend upon photosynthesis, are denied their basic existence. The length of darkness is long, perhaps six months to a year. For many plants, the effects will not be terminal. Seeds can survive many times longer than this; pieces of roots and rotting branches will survive long enough that when the light returns new growth can begin again. Most animals cannot survive without food for this long and their populations are doomed to decline to levels where extinction is perhaps possible. This scenario of impact winter is a popular one and at the moment there is good support from computer models that it could happen, particularly if fires are ignited and release soot into the atmosphere. Helped along by sulfates from rocks injected into the upper atmosphere, this concoction might be very effective at cutting out light. Apart from the putative soot layer, there isn't any unequivocal evidence in the fossil record that this ever did occur, but it is one of the favored mechanisms by which an impact event would bring destruction to surface-dwelling animals that are directly or indirectly dependent on photosynthesis.

The darkness of the impact winter would pile further stress on the biosphere. Ultimately most sea and land creatures depend on photosynthesis. Even many deep-sea fish, far away from the warmth of the Sun in the deepest parts of the oceans, depend upon dead algae that rain down from the surface of the oceans. This "marine snow", as it is often called, provides food for invertebrates, like shrimp, and thus the fish that live in darkness. To begin with they would be robust against the shut down of light, but probably perturbations in even their food webs would eventually occur. What survives this period of darkness is speculative. One might venture to speculate that organisms that would do particularly well are detritivores – organisms that can live off the dead and decaying matter of other organisms.

Beetles that eat rotting wood or rotting flesh and small mammals that eat the roots of plants are animals that might remain completely unaffected by such an event, provided, that is, they have survived the fires and the toxic chemicals. In some cases one could even imagine that their populations might explode. Imagine a mass die-off of billions of animals over the globe and the food this would provide for anything that eats the dead and decaying flesh of other animals.

A few years ago I spent a year working at Biosphere 2, a large glass structure north of Tucson, Arizona. Enclosed within this structure are artificial forests, deserts and an ocean. The site of a remarkable experiment to enclose six people for two years, the Biosphere is now owned by Columbia University and is the focus of work on rainforests and their interactions with atmospheric gases (and a diversity of other projects which are based on the ability to control the environments of these huge ecosystems). One anecdote of interest to our discussion here is what happened to the crazy ants. Crazy ants (*Paratrechina longicornis*) are found in America and are so called because of their habit of walking everywhere at great speed. They were accidentally released into the Biosphere in 1991, soon after its construction, in the soil of a plant.

Soon the crazy ants took over. With a voracious, and apparently insatiable, appetite for protein they ate other ants, then other

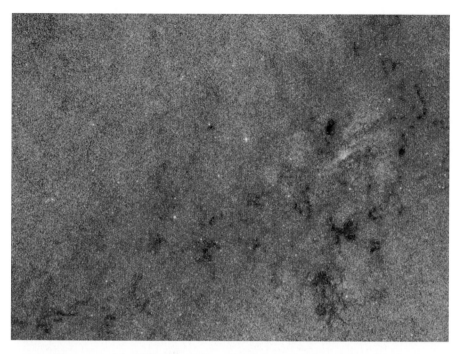

PLATE I The centre of the Milky Way Galaxy. This image was taken in the infrared and shows the centre of the Milky Way, around which we rotate in our 225 million year galactic journey.

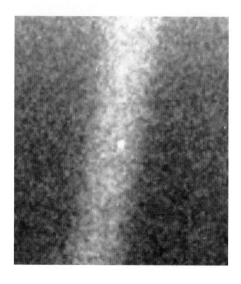

PLATE II What were you doing on February 14, 1990? The Earth as a dot photographed 3.7 billion miles above the Solar System by the spacecraft Voyager 1 on that date. A ray of light, reflected from the spacecraft's instrument package, is cast across the image. (Image, NASA.)

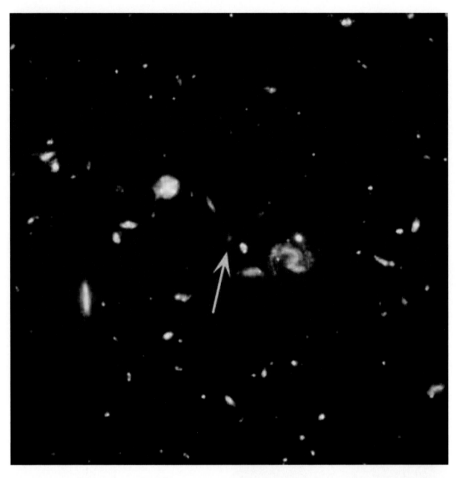

PLATE III The Hubble Deep Field View. The marked red galaxy is perhaps one of the oldest galaxies to be seen. The light from the galaxy that hit the camera, giving rise to the red speck, probably set out from the galaxy 14.5 billion years ago, 11 billion years before the evolution of life on Earth. The photograph took ten days of exposure (Image Space Telescope Science Institute/NASA.)

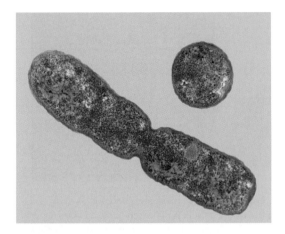

PLATE IV A picture of *Eschericia coli*, a small microbe just a thousandth of a millimetre across at the waist (also shown is a cross-section through the microbe). *E. coli* is to be found in the guts of animals, including our own. Errant forms of the bug can cause serious food poisoning (Image, Dennis Kunkel.)

PLATE V Demise of the dinosaurs? The prevailing view of the extinction of the dinosaurs, as this *Tyrannosaurus rex* might testify, involves the collision of an approximately 10 kilometre diameter asteroid with the Earth. Other factors might also have contributed to the demise of so many species at the boundary (Image, Joe Tucciarone.)

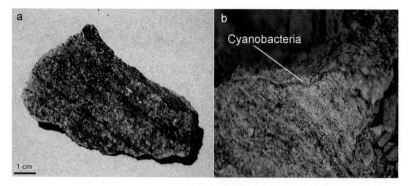

PLATE VI Opportunities for life in asteroid and comet craters. In the shocked gneiss of the Haughton crater in the High Arctic, cyanobacteria take advantage of the porous, translucent, rocks to make homes. (a) A piece of gneiss that has not been shocked by the asteroid or comet impact. (b) A layer of cyanobacteria living in a piece of gneiss that was heavily shocked by the collision in the arctic 23 million years ago. This is just one instance of how microbes can proliferate after an impact event. (Image, Charles Cockell.)

PLATE VII A harbinger of bad times to come? This red dwarf at the centre of the picture, at a distance of 63 light years in the constellation Ophiucus is approaching our Solar System and is estimated to pass within one light year from us in about one and a half million years from now. Potentially disrupting the Oort cloud around our Solar System, will it send comets into the inner reaches of the Solar System? (Image, Digitised Sky Survey, SkyView.)

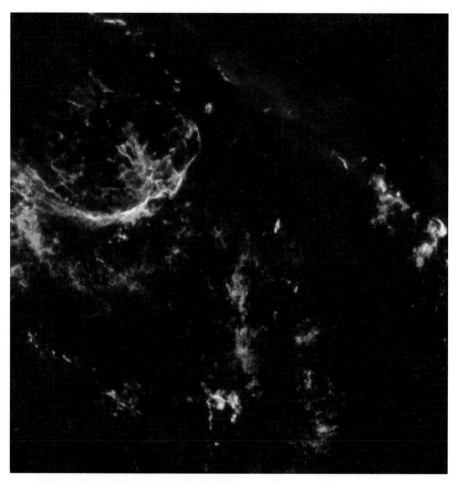

PLATE VIII Three thousand years ago a star exploded in the Large Magellanic Cloud, 169,000 light years away. Material thrown out from the supernova plows into neighboring clouds to form these luminescent shock fronts. The blue-green filaments are formed from oxygen-rich gas, showing how the oxygen that you and I breathe formed in the earliest exploding stars. The effect of such a remnant shell sweeping through our Solar System is predicted to be a long period of ozone depletion. The region shown is over fifty light years across (Image, Space Telescope Institute, NASA.)

PLATE IX Another way to survive a supernova is to live under rocks. Rocks protect from ultraviolet radiation and perhaps also mitigate the effects of some other high-energy particles. Here, cyanobacterial growth can be seen on the underside of a rock from the Canadian High Arctic. The green region was embedded in the soil. The light penetrating around the edges of the rock is just enough for this thin layer of photosynthetic organisms to be able to live underground. The surface of the rock is covered by a black layer of cyanobacteria, rich in the ultraviolet-protecting compounds that they need to survive in this habitat. The rock is about twenty centimetres (eight inches) long. (Image, Charles Cockell.)

PLATE X A pyroclastic explosion from Mount Saint Helens in Washington State, USA. The injection of ash high into the atmosphere can block out sunlight and cool the Earth. In this case the explosion was too small to do this on a large scale, but it is thought that much larger volcanoes in the past could have had global-scale effects.

PLATE XI We may think that the microbial world has never seen anything like our waste before, but deep in the Uranium mines of the African country of Gabon are fossil nuclear reactors that operated two billion years ago when the Uranium deposits were rich enough to trigger their own fission reactions. Here we can see reactor No. 15, the waste products are the area of yellow in the rock. Even waste from our nuclear weapons plants is probably old hat to the microbial world (Image, Robert Loss)

insects and then they harassed the birds and reptiles into extinction. By 1998, apart from the plants and microbes, there was little alive in the Biosphere apart from the crazy ants. Their legacy in the Biosphere is the lesson that ecosystems out of equilibrium can fall mercy to a single species with a voracious appetite and a willingness to compete. Only with protein supplies now on the wane, do the ants look like loosing their hold on the Biosphere. In the post-impact world one can expect the perturbation to result in population explosions as well as collapses. New food resources that are made available will trigger the explosion of some species, otherwise kept under control by competitors and predators, and they will have free reign until equilibrium again begins to emerge in a stable post-impact biosphere.

As the soot and dust begin to clear a new threat will await the emerging biosphere. Compounds formed by the impact, particularly nitrogen oxides, might have depleted the ultraviolet-protecting ozone layer and it may continue to be depleted even after the soot and dust have begun to clear. Levels of ultraviolet radiation could be twice as high as they were before the impact. These nitrogen oxides will rain out as acid rain into streams and lakes and could, by dissolving away organic compounds in lakes and ponds that help protect against ultraviolet radiation, help to increase the biological effects of the ultraviolet radiation. Although the ultraviolet radiation will almost certainly not itself cause extinctions, it will further add to the stress experienced in the biosphere.

The amalgamated effect of all these stresses is to bring about a collapse of the Earth's biosphere, primarily through a wholesale shut down and destruction of photosynthesis, upon which so much of the Earth's biology depends for its energy.

Some types of microbes, even in the surface regions of the Earth, might survive quite well. Groups of microbes that can photosynthesize when there is sunlight, but switch to eating other microbes in darkness (the "mixotrophs") might succeed. Perhaps the dark post-impact oceans are a thriving habitat for predatory mixotrophic species? Astonishingly there is good fossil evidence that the dinoflagellates, which are a mixotrophic group of algae, did very well at the

K/T boundary. Some speculate that they did well because they could hibernate as cysts and wait until better conditions arrived. I'd bet that they did more than merely form cysts, but actually grew in the post-impact oceans, living off their neighbors when the light had gone out. Some people have speculated they do exactly this in Antarctic lakes during the polar winter, when the continent is plunged into darkness.

Are there any impact calamities large enough to kill off even the subsurface microbes? What about an event even bigger than the supposed K/T impact event?

During the early history of life on Earth, when the rubble from the early Solar System was more abundant and, as a result, large impacts were more common, there could have been impacts so violent that even the oceans boiled. The Imbrium basin on the Moon, a 1,200-kilometer lunar dust bowl, is testament to the fact that some enormous asteroid and comet impacts have occurred before. On early Earth, as microbes were beginning to innovate and occupy different habitats, a rain of material from the Solar System may have frustrated the evolution of life. Perhaps, just under four billion years ago, the Earth could have been hit by asteroids up to 400 kilometers across, forty times bigger than the one that is supposed to have been responsible for the K/T extinctions.

Such an asteroid (Figure 13) would boil away the oceans and transform the atmosphere into one of steam and molten rock. The energy would be so great that the temperature of the atmosphere could have been well over one thousand degrees centigrade for two to three thousand years after the impact. These temperatures are well beyond the upper limit for life and the surface of the Earth would have been baked free of living microbes. Were there regions in the deep subsurface that remained below one hundred degrees centigrade? Trapped between the sterilizing heat of the Earth's core and the lethal effects of a 1000 °C atmosphere, could there have been subsurface oases cool enough to support life? Some scientists suggest the answer is "yes" and even go so far as to suggest that the reason why the microbial world, and thus life itself, seems to have evolved from

FIGURE 13. An impact on this scale would be sufficient to boil the oceans. Fortunately they are now rare (perhaps one every 2 billion years), but on early Earth they were more common. (Image, NASA.)

heat-loving microbes, the earliest common ancestor of life, is because these microbes, the so-called "hyperthermophiles", were able to survive these early ocean-boiling asteroid impacts.

It doesn't take much imagination to see that the vaporization of the oceans would probably have frustrated the evolution of life. However, the very suggestion that these impact events selected for heat-loving microbes is an admission that even under these intensely difficult early years of Earth's history, there were organisms that could survive. Microbes that can grow at high temperatures near the boiling point of water could have remained in habitats and proliferated, even whilst the very oceans around them were periodically boiled.

On early Earth impact events were so common that their frequency was much shorter than the lifetime of a microbial species. Thus, survival of impact events wasn't survival of the luckiest, but true Darwinian survival of the fittest. The survival of one impact event was because their heat-loving attributes had been selected by the previous impact event. Thus, on early Earth evolution really did occur in the cosmic context. Organisms did well because they could survive the vagaries of the galactic journey. Soon, the impact hazard would drop off as the newly formed planets sucked up the remnants of the early Solar System. The frequency of impact events with global-scale effects would drop to below one every five million years,* which is thought to be about the average lifetime of a species (although the average species lifetime for many microbial species isn't known). Now, however, there would be many organisms that weren't even around during the last impact, and their survival could only be because they fortuitously possess traits that allowed them to survive. The mode of evolution for many might have switched from survival of the fittest to survival of the luckiest.

There is more to the story of the microbial survival of asteroid and comet impacts during our galactic journey. Something far more remarkable than merely "making it through". Some microbes are presented with new opportunities. They actually find new homes in the post-impact environment.

MORE ABOUT IMPACT CRATERS

The Haughton crater (Figure 14) is only one of about 160 asteroid and comet craters that have been identified around the world ranging in size from just a few meters to hundreds of kilometers across. Scattered through every climate zone of the Earth, they range from bush-filled craters in the Australian desert to lizard-infested craters in South Africa to craters with impact lakes filled with cannibalistic trout in Canada. A remarkable diversity of ecosystems is to be found in these structures. Often the hole in the ground created by the excavation of material during the impact allows water to drain in and form lakes, but in some places, such as the Australian desert, it is so hot that the water evaporates and salt deposits build up in the crater (Figure 15). In one crater, Lappajärvi, in Finland, humans have taken advantage of the site by building a ski ramp on its rim.

* It has continued dropping and is below one every 100 million years today.

FIGURE 14. Haughton crater today. The 24-kilometer impact structure is clearly visible from space in the polar desert below. (Image, Geological Survey of Canada.)

Together with my colleagues at NASA, I have recently investigated how microbes colonize the rocks in the Haughton impact crater, a 24-kilometer diameter impact crater in the frozen polar wasteland of the Canadian High Arctic. The crater is on Devon Island, the largest uninhabited island in the world at 75° N. There is little growing here. Apart from the tiny arctic willow, cousins of willow trees just a few centimeters long, that hug the ground with their tiny wiry branches, the plant life is sparse. Flying in from the town of Resolute, an aptly named settlement of about two hundred Inuit that lies about one hundred kilometers south of Devon, the first you see of the crater is a vast, gray circular structure. During the impact of the asteroid or comet 23 million years ago many of the rocks were heated to immense temperatures, up to 10,000 degrees centigrade and subjected to pressures equivalent to many thousands of atmospheres. The equivalent energy of 100 million Hiroshima bombs was released during the collision and the crater became filled with the melted rocks like an enormous liquid-rock bath. The rocks cooled into the concrete-like gray hills that remain there today as the rather imposing melt rock hills, eroded by the Haughton River, but intact across much of the crater.

FIGURE 15. Craters are influenced by the climate in which they sit. (a) The Tswaing impact crater in South Africa, a 1.1-kilometer diameter crater in the African bushveld just north-west of Pretoria. It contains a hypersaline lake. (b) The 880-meter Wolfe crater in the deserts of East Kimberley in Australia. A saltpan occupies the middle of the crater where the water has evaporated in the desert heat. (Image, Charles Cockell.)

The crater is a dusty place, a little like the dust-ridden surface of Mars. Each summer since 1997, a small village of NASA scientists has called the dusty plains on the rim of the crater their home. The white tents, enclosing computer technology, robots, a basic kitchen and vehicle spare parts, become surrounded by the patchwork of many individual sleeping tents belonging to the scientists, in oranges, reds, blues and greens. We mainly go there to test technologies for the human exploration of Mars. It is a project called the NASA Haughton–Mars Project, set up by Pascal Lee at the NASA Ames Research Center in California. In a collaboration with the Mars Society, a simulated Martian habitat eight meters in diameter has been born in the crater and each summer it is used to test protocols for communications and mobility on Mars. The reason for this work is that Haughton is situated in the Arctic permafrost and Mars is a planet covered by permafrost. Because there are no plate tectonics on Mars, the surface, which has been pummelled by asteroids and comets for 3.5 billion years, still preserves a large record of impact cratering. Haughton crater is in some respects a reasonable analogue for the impact craters on Mars and this is why is has become a focus for testing the science and technologies needed for humans to establish a permanent presence on Mars.

I digress, because my reason for being in this crater was to find out how microbes can take advantage of a place that was transformed by a destructive asteroid or comet impact. In amongst the gray hills of Haughton are lumps of a black crystalline rock, called gneiss, brought up from one and a half kilometers under the ground by the violent excavation caused by the collision. During their excavation, the lumps of gneiss underwent great changes. Not only did some of their mineral components vaporize, but many fractures and cracks were formed in the rocks. The result of these physical changes would have profound consequences for microbes. The cracks and fissures make it easier for microbes to get into the rocks and make homes inside them. The vaporization of many of the minerals made the rocks white and translucent, compared to the dark, opaque gneiss that was not shocked by the impact event. The translucence allows through light for microbes to gather energy.

Today there are photosynthetic cyanobacteria, called *Chroococcidiopsis,* inside the shocked rocks. They are far more abundant there than in the unshocked rocks. Because of the translucence of the rock, the cyanobacteria are able to get enough light to gather their energy through photosynthesis. In Plate VI you can see an example of a layer of green cyanobacteria inside one of the shocked rocks from Haughton and next to it is a blue-green gneiss that was not as heavily shocked by the impact event. So here is an example of a habitat that was actually created by the asteroid impact, a habitat that was not there before. Out of the violence of an impact event the microbial world has found a new opportunity to proliferate itself.

The shocked rocks weren't alone in providing a habitat for life in the newly formed crater. The energy delivered by the impact created steaming hydrothermal vents. The heated minerals that attest to these hot-water vents can still be found today and by examining these it has been proposed that the vents might have provided hot water habitats for heat-loving microbes for 1000 years and maybe up to 10,000 years after the impact.

As the vents spewed hot water into the crater it began to fill. A lake formed within the rim of the crater and within it microbes could bloom. Over thousands of years sediments built up at the bottom of the lake and preserved the plants and animals that fell into it. The sediments of Haughton today preserve the only record of the Arctic over 20 million years ago. Giant rabbits and woolly rhinoceros roamed the forests around the crater, perhaps using the crater lake as a watering hole. No doubt birds passing over this region of the Arctic stopped to drink. For tired, wandering animals migrating through the pine forests in spring and autumn, the lake would have provided a welcome break.

It isn't known when it happened exactly, but the Haughton River finally breached the rim of the crater and the lake drained away, leaving the sediments exposed. They eroded away. Today only a seven-kilometer square patch of sediments remains in the crater. They were cored in the 1980s with drills to see what fell into the lake when it still existed. The cores gave us the insight we have today of what lived near the crater after the impact.

Today the lake sediments offer new opportunities for life, long after the rabbits and rhinoceroids went extinct and the forests retreated south. The rich exposed sediments are covered in arctic grasses; from a Twin Otter aircraft bound for the crater they hit you as a wonderful and luxuriant expanse of green against the lifeless gray and brown of the polar desert. The grass is food to many animals, including a family of arctic fox that has settled here. Standing on the edge of the oasis, you can watch the mother leading her two cubs in search of food, their gray coats wonderfully like the gray hills behind them. They make a fine picture as they play around their den, the young ones nibbling at each others' ears and, every now and then, running to their mother for safety. They are blissfully unaware of the fact that their lives were made possible by the collision of an asteroid or comet with the Earth that destroyed life for hundreds of kilometers across the Arctic millions of years ago. A family of woolly musk ox graze the lush vegetation and arctic owls sit in their nests on frozen mounds of vegetation, launching themselves into the air to hunt the lemmings that burrow and graze in this oasis. Twenty-three million years later the sediments that resulted from an event of catastrophic destruction provide abundant life.

Haughton probably wasn't a global killer. The scale of the event probably confined its affects to the "local" environment. When I say local of course, it might still have extirpated life for several hundred kilometers around the crater. Within this crater is evidence that novel habitats for life were created by the impact event; habitats that were not there before the collision. The crater shows us that not only will some organisms survive impact, but they will positively benefit. This is particularly true of the microbes. Most of the opportunities created at Haughton, including the hydrothermal vents and shocked rocks, were of benefit to the microbial world.

Many of the global environmental effects of large impact events remain speculative. Although the debate rages about whether an asteroid killed the dinosaurs, the fact that very large impacts can potentially cause extinctions is undoubted. What is also undoubted is that for many of these events, the microbes that live five kilometers under the ground will be unscathed. Having energy supplies, as they do, that are quite independent of processes on the surface of the Earth, they will make it through most impact calamities, inhabiting a subsurface refuge from impact-induced extinctions. And as we have just seen at Haughton, opportunities can be created for microbes as well. These new chances for life inside a crater would also exist in the much bigger structures associated with global-scale extinctions.

Just as survival of impact events today is probably about luck for most microbial populations, so taking advantage of the new environments create by them is all about opportunism. The cyanobacteria that colonize the shocked rocks in Haughton can be found all over the Arctic. There is nothing special about them; they are even found in hot deserts like the Negev in Israel and in the Antarctic too. They are opportunists though; they will colonize the rocks in the crater because they are colonizable, not because they have special attributes that came from their ability to previously colonize impact craters.

Impacts, to use a terrible cliché, are a double-edged sword. The long interval between globally destroying impact events makes it possible for complex ecosystems to evolve and develop before they are next destroyed. Perhaps this increases biological diversity. However, on the other hand, the long interval makes life more vulnerable to impact events by allowing the evolution of a complex biosphere that is, for the most part, lacking in the selected traits to survive impacts. Inter-impact evolutionary decadence sets in as animals evolve that are poorly equipped for the statistical certainty of impact events. As we have now seen, it is the microbial world that has bucked the trend. Although not all will survive these events, many are generalist enough to ride-out these calamities and many are opportunistic enough to rapidly take advantage of the new environments that are created (Plate VII). The meek, apparently, are likely to inherit the Earth.

6   Supernova fry up

For life, the cosmos is a strange irony. Violent events that begat life are also those that threaten it. Supernova explosions are the origin of the elements from which life began and yet their violence makes them objects best observed from a safe distance. But just what is a safe distance? And what happens if you happen to be closer than the safe distance? Astronomers with an interest in biology began to wonder about these things almost as soon as they had discovered the frequent nature of this interstellar violence.

Supernovas, as I described in Chapter 1, are the explosions caused by the collapse of massive stars getting to the end of their adult lives. The heated outer layers give off an intense burst of ultra-violet radiation and visible light that slowly disappears over a month or so. This is the supernova light curve that astronomers are keen to catch early because this light tells them much about what is going on in the center of the dying star.

The biological effects of supernovas are even more speculative than those of asteroid and comet impact events. Like the fact that humans have never observed a large-scale impact event, we have never observed a supernova explosion close enough to influence life on Earth. And unlike impacts, they are very difficult to find in the fossil record. The signs of their past influence on life are at the moment quite elusive.

Calculations suggest that sometime during our galactic journey the Earth has a good chance of being close to one of these explosions. If one takes an average rate of supernovas occurring in our Galaxy as being once every ten to a hundred years, then we can calculate how often we would expect one to go off at about 35 light years away or closer. At this distance damage to the ultraviolet-protecting ozone layer could occur (Plate VIII). There is probably one explosion close enough to affect our planet every 70 to 500 million years. In other words during our 225 million year journey around the Galaxy, there is a good chance that the Earth will be affected by one of these events. And I should add that they aren't the only sources of radiation. Neutron stars – stars in which the matter is so dense that it is made up purely of neutrons, can collapse into each other in binary star systems and generate large quantities of radiation. At a distance of a few thousand light years, these events could also cause ozone depletion.

It was Mal Ruderman, at Columbia University in 1974, who was the first to calculate what might happen to the Earth if a supernova explosion occurred near by. He recognized that the cosmic rays from the supernova explosion, that he calculated could last for up to three centuries, would cause the formation of nitrogen oxides in the high atmosphere. They would deplete the ultraviolet-radiation protecting ozone layer.

The most damaging type of radiation that reaches the surface of the Earth is ultraviolet radiation as anyone who has burnt themselves on a summer vacation will know. As its name, "ultra" or "beyond" suggests, it is below violet, the part of the spectrum invisible to human eyes. When this high-energy light is absorbed by molecules of importance in biology, like the genetic material DNA, it can cause damage. The genetic material is particularly important because this is the hereditary information, that the molecule passes from generation to generation. If it gets damaged, then a mutation or error could kill the cell. If you are a large animal and you have trillions of cells, damage to one cell that kills it may not be too bad (unless instead of dying it develops an error and turns into a cancer – a concern amongst sun-bathers). However, you can understand that for the humble microbes, made up of a single cell, irreversible damage to the DNA could be lethal and hence the importance of ultraviolet radiation to the microbial world.

Most of the ultraviolet radiation emitted by the Sun is screened and scattered by the atmosphere. A particularly important screen is the gas, ozone. Formed by the action of ultraviolet radiation itself on oxygen in the atmosphere, ozone collects in the upper atmosphere. The oxygen molecule, which you and I need to breathe, is made up of two oxygen atoms bound together. Ozone is made up of three of these atoms bound together, so in some ways it is a type of oxygen. The concentration of ozone in the atmosphere isn't very great. If you took all the ozone and compressed it onto the surface of the Earth it would form a layer merely three millimeters thick. If this thin layer of ozone is destroyed it allows shorter wavelengths of ultraviolet radiation through and the damage to biological systems is much greater. This is why people are concerned about the "ozone hole" that forms over the Antarctic in the spring that has been caused by the actions of chemicals produced from human industry.

To begin with, a supernova explosion might actually cause the ozone layer to get thicker. The burst of ultraviolet radiation from the star in the first month or so would contain the very short wavelengths that are involved in forming ozone. So Earth could experience a period

MORE ABOUT THE OZONE HOLE

The ozone "hole" is a rapidly changing shape. Many imagine it to be a round hole, but it is usually oblong and it shifts and changes shape depending on where the circumpolar vortex over the Antarctic concentrates the ozone-depleting chlorine compounds in the stratosphere. It forms in spring when there is light available to drive the reactions, but temperatures in the high atmosphere are cold enough to allow the reactions to occur at the right rate. Although human compounds like chlorofluorohydrocarbons (CFCs) are responsible for ozone depletion now, it is probable that the ozone layer has been depleted before. Supernova explosions are one mechanism, but asteroid and comet impacts might deplete the ozone layer as well and maybe even volcanoes, by producing ozone-depleting chlorine compounds. It is thought that bans on CFCs will cause a reduction in the hole by the twenty-second century, but this depends on all countries complying with the regulations.

with an ozone "blanket" when ultraviolet radiation would be reduced, rather than a "hole", although the effects of this on life on Earth would probably be insignificant because other things, like clouds, routinely cut down ultraviolet radiation. Once this burst has died down, the biggest threat comes from the stream of cosmic rays that emanate from the site of the explosion and fan outwards at the speed of light in all directions. These are the particles that would cause the depletion of the ozone layer.

The arguments about the extent of this depletion have been vigorous. John Ellis at the European Organization for Nuclear Research and David Schramm from the University of Chicago argued, using models of the ozone layer, that it could be depleted by 95%. They suggested that the ultraviolet radiation would increase to such levels that a mass extinction would be caused. They did correctly recognize that the most important effect of this elevated ultraviolet radiation would be to stress photosynthetic organisms, the base of the world's food chains, which would have knock-on effects throughout the biosphere. However, their calculations have been challenged. Paul Crutzen and Christoph Bruhl at the Max Planck Institute for Chemistry in Germany did similar calculations and came up with a different number. They suggested that a supernova explosion close to our Solar System, at about 35 light years distance, would deplete the ozone hole over the poles by about two-thirds and over the equator it would be reduced by only about a quarter. What is interesting about their calculation of two-thirds ozone depletion is that this is not far off what humans have already caused over the Antarctic with the springtime ozone hole. There are no mass extinctions in Antarctica. So one can see that the potential for supernovas to cause extinctions by ozone depletion is very much dependent on the amount of ozone that is lost.

Every spring, during September through to November, an ozone hole opens up over Antarctica and there has been recent concern that there may be one over the Arctic as well. During the Antarctic spring, concentrations of ozone drop to about half of what they are near to the hole and sometimes even lower. In other words Antarctica in spring is a rather good analog for what would happen during a supernova explosion. The difference between the ozone hole in the southern hemisphere and the one caused by a supernova is that the cosmic ray emissions from a supernova could happen for a long time, so that the ozone loss would be continuous over decades and some, like Ruderman, have even predicted centuries. Human-caused ozone depletion is, we hope, a relatively temporary affair, particularly now that bans on the chemicals that cause ozone depletion, CFCs, are beginning to take effect. Nevertheless, we can look at how organisms deal with ultraviolet radiation, both in the polar regions and in other regions of the world to see if the effects of a supernova explosion would be very dramatic.

Life has a wonderful array of ways to deal with ultraviolet radiation and they all boil down to some quite basic principles. To reduce the damage that ultraviolet radiation causes you can do a series of things in a logical, sequential order. Firstly, you can escape it altogether, then, if that isn't practical, you can screen out the ultraviolet radiation before it gets to biologically important targets. This is important for organisms that photosynthesize. They have the paradox that they need visible light (the light that you and I can see) to get the energy they need, but by exposing themselves to visible light they get a dose of ultraviolet light. For them, simply escaping ultraviolet radiation isn't possible. They use natural ultraviolet-screening compounds. They are natural versions of the suncreams so familiar to sun-bathers. Finally, if the screening isn't a hundred percent efficient, they can repair the damage caused by the ultraviolet radiation.

For some, escaping ultraviolet radiation is just a matter of getting into darkness, for instance into caves and underground burrows. Nocturnal bats and rodents have no worries about radiation damage. They almost certainly didn't take up the habit of being nocturnal to escape ultraviolet radiation; there are other more important selection pressures for being nocturnal, but the nocturnal life fortuitously happens to be a lifestyle that is free of the problem of radiation damage.

Microbes that live in the ponds and water systems of caves also get the advantage of ultraviolet radiation protection as do all microbes that live in the deep subsurface. The microbes in the deep Antarctic ice and in the deep igneous rocks of the Earth that we saw in Chapter 3 will be unaffected by any changes of ultraviolet radiation on the surface of the Earth.

Even some animals that live out on the surface of the Earth under the warmth of the Sun can be shielded from ultraviolet radiation. Thick layers of fur or feathers block it out. The bodies of most furry animals, from cats to polar bears, are protected. These animals do have one Achilles heel though – their eyes. Eyes require light for visual information and, of course, with the visible light comes the ultraviolet radiation. Cataracts can form in animal eyes after long-term over-exposure and cataracts can potentially cause blindness. The production of a variety of ultraviolet-absorbing chemicals in eyes, including derivatives of the amino acid, tryptophan, is one way in which animals can reduce the damage.

For some organisms, escaping ultraviolet radiation isn't that simple. They need light for photosynthesis and so they have to find a way of collecting as much visible light as possible for their energy needs, but reducing their exposure to ultraviolet radiation. Some clever responses have evolved.

There is a group of microbes, dinoflagellates, that have two tails that allow them to swim (they are not the only microbes that have the power of locomotion though). When these mobile microbes are exposed to ultraviolet radiation they move away from the light source to escape the damage. These microbes possess sensors that produce a signal that tells the microbe that it is being exposed to the ultraviolet radiation and the direction from which the light source is coming. By using this system the microbes can move up and down in a lake, regulating their exposure to radiation. Often some species will come to the surface of a lake in the night then during the morning as the Sun rises and ultraviolet radiation levels rise, they will journey down into the lake to escape.

This same process also occurs on a micro scale as well. Within a microbial mat just one or two centimeters thick, cyanobacteria can move up and down as they are exposed to ultraviolet radiation. This mobility allows them to organize a trade-off between the amount of light they have for photosynthesis and the dose of ultraviolet radiation they receive that will cause damage. Too high up in the mat and they get lots of light for their energy needs, but the ultraviolet radiation can be very damaging. Too low in the mat and they effectively escape the ultraviolet radiation, but the visible light levels are too low for them to get energy. Somewhere between these two extremes is a balance. The power of locomotion allows microbes to get this balance just right at different times of the day.

Running away is certainly quite effective, but there are ways in which you can simply remove the ultraviolet radiation yourself before it causes damage.

All of us are familiar with what happens when we go on a summer vacation. We tan. Tanning is a cunning method of protecting ourselves from radiation. The brown coloration that builds up in our skin is caused by the production of melanins, which are complex molecules that absorb ultraviolet radiation. The brown color is caused by the fact that they also absorb some visible light, so you can actually see a brown color change occurring. For an animal like ourselves that lacks hair on a lot of our body, being able to synthesize compounds that protect us from ultraviolet radiation is very important. If damage to skin cells was to result in fatal cancer, then you could die before breeding age, failing to pass on your genetic material, so from an evolutionary point of view it is essential to have ways to reduce the damage.

Screening compounds can perform yet more subtle tricks. Plants need to be exposed to light to photosynthesize, like some microbes. Unfortunately, unlike some microbes that have locomotion, they can't move away from radiation very easily. Producing a brown melanin-like compound that blocked out ultraviolet radiation, but also blocked out visible light, so essential for gathering energy, would not be very useful. A selection pressure has therefore been imposed on plants, whereby the most successful plants are ones that can screen out ultraviolet radiation, but allow through as much visible light as possible for gathering energy. The answer to this challenge has been the evolution of a range of molecules, including the flavonoids, that have chemical structures that absorb in the ultraviolet radiation range of the spectrum, but are transparent in the visible region. If you extract flavonoids from plants using a chemical such as methanol, you end up with a clear solution that looks like water. Put this solution into a machine that will measure the absorbance in the ultraviolet radiation region of the spectrum and you will see that these compounds are blocking the ultraviolet radiation. Flavonoids are produced in the leaves and some other organs of plants. They are, if you like, sun creams for plants.

The screens are not one hundred percent efficient and a variety of effects of high ultraviolet radiation on plants have been reported. Some plants get stunted, some become less good at photosynthesis and some change leaf size. These changes in the way they grow may in turn change the competitiveness they have against other plants. However, few plants are killed outright by elevated ultraviolet radiation, even by up to a simulated depletion of a third of the ozone layer. Nevertheless, research shows that even with all these screening compounds, ultraviolet radiation is a quite important regulator of the way in which plants grow in nature.

Could an animal or plant ever be made to go extinct by an increase in ultraviolet radiation? It is difficult to see that it could directly make animals or plants extinct because of all the protection mechanisms they have but, by changing competition between species, it is conceivable. If one species of plant was more sensitive to ultraviolet radiation than another, one could image that an increase in ultraviolet radiation would give an advantage to the less sensitive species. Perhaps this would be enough of an advantage that it could outgrow the other species, eventually driving it to extinction, by reducing its numbers to below the minimum viable population. This remains speculation.

This book is about microbes, and although they are smaller and simpler than plants or humans, they have been no less cunning in their use of ultraviolet-radiation-screening compounds. Can we get some insights into how they might fare against ozone depletion caused by supernovas?

Some of the most primitive photosynthetic microbes to appear on Earth, the cyanobacteria, produce a compound called scytonemin. A brown-orange compound, it screens out short-wavelength ultraviolet radiation, but allows enough visible light through for photosynthesis. You can see a typical microbial mat that produces copious quantities of the compound in Figure 16. The uppermost layers of the mat are packed with scytonemin and within less than a millimeter into the mat all of the ultraviolet radiation is completely screened out. Mats like this one grow in Antarctica.

One of my favorite places to see these is in a little ice-free spot called "Mars Oasis" next to Utopia Glacier on the Antarctic Peninsula. The Oasis lies next to impressive sandstone terraces that rise next to great ice ridges caused by the crashing of the ice sheet against land. You reach it by flying in a twin otter aircraft, the workhorse of polar regions, from the British Rothera Research Station about two hundred kilometers north. Sitting on the edge of Alexander Island facing King George VI Sound, the Oasis is, as its name suggests, a small location where life can thrive in an otherwise frozen continent at 72° S. The Oasis thaws out during summer and small bodies of water form in which communities of cyanobacteria can flourish. Even under an ozone hole, the organisms living in the mat are shielded from ultraviolet radiation, but enough visible light probably penetrates for them to be able to gather their energy through photosynthesis and grow.

Few of these clever defenses are single compounds doing all the work. The microbes that we saw in the Antarctic mats, cyanobacteria, produce a suite of colorless compounds, which absorb ultraviolet radiation, called mycosporine-like amino acids or, more easily remembered, "MAAs". Unlike scytonemin, the MAAs are not confined to the cyanobacteria. They seem to have been an early evolutionary

FIGURE 16. The supposed ozone depletion and thus increase in ultraviolet radiation that would be caused by the explosion of a supernova close to the Earth could be survived by many organisms. Here we see the thick layers of a microbial mat. The upper layers protect the lower layers in the mat from ultraviolet radiation as they are packed with the cyanobacterial ultraviolet-screening compound, "scytonemin". These types of mats survive happily under the Antarctic ozone hole. (Image, Charles Cockell.)

innovation because they are found in many different types of algae as well. The compounds come in many different types. All of them share a common type of chemical core structure, but by making subtle changes in the chains attached to the core structures they can be made to absorb in different regions of the spectrum. So by producing a range of MAAs, microorganisms are able to screen across most of the ultraviolet region of the spectrum, preventing damage.

No-one really knows how ultraviolet-screening compounds first evolved. They may have started off being used for other purposes and then became screening compounds. The MAAs are produced by some cyanobacteria as chemicals that help them deal with drying out or excess salt. One can imagine that on early Earth they might have been produced to help prevent desiccation and then they fortuitously provided some protection against ultraviolet radiation as well. Eventually they could have become an important response to ultraviolet radiation in their own right. The same type of scenario is proposed for the screening compounds in plants. It is imagined that the first flavonoids might have been chemical messengers in plants, or maybe they were produced to stop predators. Because their structure absorbed ultraviolet radiation, any plants that had them might have been better protected against ultraviolet radiation.

Although organisms can produce their own ultraviolet-screening compounds, one particularly fascinating discovery is that predators can ingest MAAs found in their prey and this can give them some protection. Sea urchins that graze off algae containing the compounds seem to collect MAAs in their tissues and these compounds protect them and their young.

Although evolution has provided some quite clever screening compounds there are some more simple ways of protecting yourself – and that is to use a screen that is already available in the environment around you. Undoubtedly the most common is rock. Some microbes, particularly the cyanobacteria, live within the cracks in rocks and simply take advantage of openings in the rocks to invade the cracks. Others manage to get into more unusual places in the rock. By invading the surface they get into the inter-grain spaces in the rock structure and grow within the rock, living just a few millimeters below the surface. Because they need just the right spacing between the grains of rock to be able to invade, they are more selective about the types of rocks they grow in than the microbes in cracks that can, of course, take advantage of the first crack they find. These rock-dwelling microbes are very similar to the ones I described in the last chapter, that live in the rocks in the Haughton impact crater in the High Arctic and you can see an example of microbes living in rocks in Plate VI. Inside these rocks the microbes are protected from ultraviolet radiation.

In the dry valleys of the Ross Desert in Antarctica these microbes invade sandstones and form a community with different layers, like a tiny layered cake. One zone is formed by algae, another by fungi and a third by lichens. In this way, some quite complex communities can build up inside the rock. In Plate IX you can see yet another rock habitat to live in; the underside of the rock. The microbes can grow on the bottom of the rock, using the whole rock itself as protection from the ultraviolet radiation above. The habit of living inside or under rocks isn't restricted to cold polar deserts. Rock-dwelling microbes are also found in hot deserts like the Negev, living in the cracks and subsurface of quartz stones.

The rocks are not just an ultraviolet-radiation screen, they also provide a miniature microclimate for the organisms. Fluctuating temperature extremes and freeze–thaw cycles can be mitigated. By trapping moisture the rocks provide more water for the microbes than the surfaces of the rocks, which the wind can blow-dry very quickly.

MORE ABOUT LIVING IN ROCKS

Planets are made of rocks and microbes can live in rocks. You can't get more basic than that as a habitat. The "lithic" habitat is perhaps one of the most widely available habitats on Earth. Microbes that live in cracks are called "chasmolithic" microbes and microbes that live in the subsurface layers of rocks (in the inter-grain spaces) are called "endolithic" microbes. Those that choose to take up residency on the underside of rocks are called "hypoliths". These habitats are shielded from many of the environmental extremes to be found in the exposed areas around the rock. Microbes can grow on the surface of rocks as well, in which case they are called "epiliths", but wind tends to scour the surface of rocks. In extreme environments like hot and cold deserts, the surfaces of rocks are often bare in all but the most shielded locations.

Of the hardiest microbes to inhabit these rock environments, probably the prize should go to *Chroococcidiopsis*, a desiccation-resistant, radiation-resistant survivor. These tiny, spherical, green cyanobacteria can lie dormant in the rocks if they have to, perhaps for centuries or longer, and wait for water to arrive.

Microbes are not merely limited to rocks. Salt flats also harbor cyanobacteria and algae where the water has evaporated and widespread encrustations of salt remain. A good place to find these encrustations is in the inter-tidal salt flats in Baja, California in the USA. The organisms which, like the endoliths, form bands of communities inside the encrustations, are halophilic, that is, tolerant to very high salt concentrations. By living inside the salt, the organisms are protected against the extremes of radiation outside.

Not all the regions of the world have pleasant balmy climates and if you happen to live in freezing glaciated areas like the Arctic or the Antarctic then other ways to escape ultraviolet radiation are on hand. Snow and ice can do a reasonable job of helping to cut down radiation. Snow scatters light. Under a snow cover of about five centimeters or more, the ultraviolet radiation is reduced to a tenth of what you would find on uncovered ground – and there can still be liquid water available for microbes to grow. Even at the edge of the snow pack, overhangs form like little roofs. These overhangs form by the melting of the snow on the ground, leaving the projecting overhang above the ground. They can be heavily encrusted by snow and air bubbles, which improve their ability to cut out the ultraviolet radiation. Some can reduce the ultraviolet radiation by up to five or ten times, providing a shielded little habitat where water is available for growth and there is enough visible light for photosynthesis.

In deep polar lakes (Figure 17), which are covered by ice all year round, the microbes are even better protected from ultraviolet radiation. Under some of the perennially ice-covered lakes of the Antarctic near the US McMurdo base, the ice can be a meter or more thick. These lakes, such as Lakes Vanda and Frixell in the Ross Desert, are an oasis for life in this desert region of the Antarctic. There is no

FIGURE 17. Ice and snow reduce ultraviolet radiation and can help absorb cosmic rays and other high-energy particles. Microbes protected under the ice-covered regions of the world, like this location on the Antarctic peninsula, might be unscathed by any near-by astrophysical violence. (Image, Charles Cockell.)

rainfall and the only other inhabitants are the endolithic cyanobacteria inside the rocks. At the bottom of these lakes there is just enough visible light for microbial mats to make a living from photosynthesis. These permanently ice-covered lakes would help protect their enclosed biota against the radiation assault that might be caused by the explosion of a near-by star.

The problem with screening compounds and living under substrates like rocks, salt and snow, is that they are rarely one hundred percent efficient. This is particularly true of the biological ultraviolet-screening compounds. It would take a lot of compound to completely eliminate all the ultraviolet radiation penetrating into a cell. So even with these compounds, some ultraviolet radiation will make it through and cause damage. This is where the next line of defense comes in. Organisms can repair damage caused by ultraviolet radiation.

Some of the damage comes in the form of free radicals, which are reactive types of oxygen. The ultraviolet radiation can react with water to form these radicals that then react with biologically important molecules like DNA. So a good line of defense is to have compounds that can mop up these reactive types of oxygen and compounds called "carotenoids" perform this job. These compounds are colorful compounds in reds, oranges, yellows and a variety of shades in between. The compound, beta-carotene, which makes your carrots an orange color is a type of carotenoid. If you take a typical soil sample and culture the microbes contained within it, you would find many different colors of microbes would grow, with spots of reds, oranges and yellows corresponding to different species that produce different types of carotenoids. By mopping up the reactive radicals, these carotenoids are essentially nipping the damage in the bud, preventing it from getting out of hand.

If an important molecule directly absorbs the energy from ultraviolet radiation, it can still get damaged regardless of whether there are any carotenoids around. The genetic material, DNA is the most important of these. The molecule is a long thread of, well, let's call them beads (actually each "bead" is a molecule), of four different types. The sequence in which these four different molecules or beads are all strung together is the code, which is translated by the cell. A single incorrect bead can cause the cell to produce the wrong protein or perhaps stall the whole reading apparatus, so it is very important that the integrity of the long code is kept intact. The cell goes to special lengths to preserve its DNA from damage. Ultraviolet radiation causes certain beads to get stuck together and when the reading apparatus reaches these stuck-together beads it can't tell them apart and it either stalls or reads the code incorrectly. The cell can die.

Organisms have an impressive method to deal with these errant beads. They can produce a protein called "photolyase" which is able to split the beads in two, reverting them back to their original individual selves and returning the integrity of the genetic code. Photolyase is switched on by light, which is quite ingenious, because it is when there is light that organisms are most likely to be damaged by ultraviolet radiation. The protein is produced by a colorful variety of organisms from mice to zebrafish, including many microbes, and it is one of the main mechanisms of dealing with ultraviolet-caused damage. Because it is found in all branches of the microbial world it is believed to be a very old system, dating back to the earliest microbes on Earth.

With any type of biochemical system it is always good to have a back-up or a few other cards up your sleeve. There are DNA repair mechanisms that cut out whole sections of DNA and re-synthesize them. Sensors that cause the production of more protein if proteins are damaged can help mitigate the problem of ultraviolet-radiation damage.

This suite of responses, from escape, to screening and repair is part of an overall response strategy to ultraviolet radiation. None of them on their own is necessarily the optimum approach to dealing with ultraviolet radiation, but by juggling different approaches organisms can survive the ultraviolet radiation they experience in the environment.

These responses are so effective that it is questionable whether a large supernova explosion causing the depletion of two-thirds of the ozone layer would affect certain groups of microbes at all. Our microbes living in mats, like those in the Antarctic, might not be affected and the microbes I talked about living within rocks and salt might escape unscathed. Under snow and ice there could be some detrimental effects, but much less than those for fully exposed microbes. The snow and ice-pack of the polar regions would be a good place to be during a supernova explosion. And as we saw above, even without these effective screens, some microbes can simply ramp up the efficacy of their repair processes and deal with the increase in the ultraviolet damage by reversing it.

Many microbes would almost certainly be affected by a sudden increase in ultraviolet radiation. Algae living in the open ocean where the waters are clear and ultraviolet radiation can penetrate to great depths might be worse affected than our thick microbial mats. Experiments in the Southern Ocean near the Antarctic continent showed that increases in ultraviolet radiation caused by the ozone hole could reduce photosynthetic productivity amongst the algae living there. Research on the effects on coral reefs suggests that the algae that make up the reefs can also be adversely affected by ultraviolet radiation. Although in some cases the decline of the world's coral reefs seems to be caused more by changes in temperature, elevated ultraviolet radiation seems to be adding to the stress already experienced by these communities.

So one could imagine that a long and continuous increase in ultraviolet radiation might initially have negative affects on global photosynthesis, which would reduce food supply for organisms higher in the food webs. How the biosphere would react over long periods to the proposed multi-century increase in ultraviolet radiation that might follow a near-by supernova explosion is not known.

Some even speculate that a supernova explosion could do some good! Because ultraviolet radiation can potentially cause mutations it could perhaps increase the number of variants in a population of a particular organism on which environmental pressures can act to drive evolution. So could elevated ultraviolet radiation be a motor for evolution, increasing mutation rates and increasing either the rate of change of organisms or the diversity of life? Perhaps supernova explosions are followed by periods of rapid evolutionary innovation. In reality these ideas are quite difficult to test. As I have already shown you, an increase in ultraviolet radiation does not necessarily cause an increase in mutation rates. Escape, screening, repair and a diversity of other processes get to work to reduce damage to DNA. Although it is known for certain that short-term increases in ultraviolet radiation can increase damage to DNA in a diversity of microbes, over time organisms can react by synthesizing more screening compounds or more repair enzymes.

We can continue to speculate and wonder what a supernova explosion would do to life on Earth, but from the point of view of the subject of this book, some facts become clear as we think about what ultraviolet radiation does to a biosphere. Many microbial communities would be completely unaffected by such an event. Deep-subsurface microbes will remain unscathed and probably a diversity of surface-dwelling microbes as well, particularly those with protective coverings. Anything that can gather energy supplies independently of being exposed to sunlight will be largely unaffected by the exploding star. And these same conclusions hold true for the other side effects that people have suggested for near-by exploding stars. It has been suggested that showers of subatomic muons from the upper atmosphere and other nasties will rain down on the Earth. Although more penetrating than ultraviolet radiation, these subatomic particles will not reach the microbes five kilometers underground, not even to one kilometer. The microbial world is safe from most of the calamities that researchers can dream of emanating from exploding stars.

As to whether evidence for a supernova could ever be found in the fossils, scientists don't really know at the moment. John Ellis and his co-workers suggested that the element beryllium might be used as a diagnostic tool for supernovas as it is an expected by-product of the explosion. They speculate that the spikes of beryllium found in the ice cores from Vostok, Antarctica for 35 and 60 thousand years ago might be evidence for near-by supernova explosions depositing this element on the Earth. They also investigated a range of isotopes of iron, chlorine and other elements as possible signatures of supernova explosions. Although unresolved, their work provides the tantalizing possibility that there may be signs of supernova remnants in the fossil record.

So far no mass extinction has been attributed to a supernova explosion. All of the five major extinctions during the last 600 million years have other more compelling catastrophes associated with them. In some ways this is not surprising. Although some have suggested that ultraviolet radiation could cause a mass extinction, I doubt it for the reasons that I have discussed in this chapter. The estimates of Crutzen and Bruhl of a two-thirds depletion of ozone over the poles are nearly the same as are currently experienced under the spring Antarctic ozone hole, which is caused by human industrial activity. This actually matches up quite well with the lack of evidence for a supernova-induced mass extinction during the last 600 million years. For the microbial world supernova explosions are perhaps one of the least threatening calamities during the galactic journey.

7   Fire from below

The harbingers of extinction do not just come from the cosmos. The Earth, with its restless molten mantle and core, has much to answer for. Punctuated through the fossil record are the signs of massive volcanoes, caused by the eruption of the liquid magma from deep within the Earth. They have contributed to life's catastrophes during our galactic journey. Volcanoes are not so obviously linked to our galactic journey as close-supernova explosions are. The latter are dependent on how close the Earth is to exploding stars, which depends upon where the Earth is in the Galaxy.

Volcanoes are, though, very much a product of the way the Earth condensed during the formation of the Solar System, a lasting biologically important legacy of the geological origins of the Earth and the Galaxy. There are two other reasons why I want to tell you about volcanoes and how they may have brought destruction to life to Earth. The first is that the environmental changes they cause are likely to be similar to those caused by asteroid and comet impacts. Volcanic winters are imagined to be similar to impact winters, caused by the injection of huge quantities of dust and ash into the atmosphere. So my discussion of our journey once round the Galaxy would be incomplete if I didn't tell you about volcanoes as well as impact events, particularly as there is such a controversy about whether some extinctions to be found in the fossil record are actually caused by volcanoes rather than impacts. The second good reason for talking about volcanoes is that large asteroid and comet impact events have been claimed to cause volcanoes. An impact, it has been said, could punch a hole into the Earth causing magma to erupt to the surface. The evidence for volcanic eruptions being caused by asteroid and comet impacts isn't very strong, but the fact that this suggestion has been a subject of debate makes it worthwhile for me to tell you about volcanoes.

Throughout our Solar System impressive volcanoes are to be found and it's really not too surprising when you consider some basic physical facts. Some planets haven't had time to cool down to the point where the interior of the planet is completely solid. Radioactive elements inside planets; particularly uranium, potassium and thorium, decay over billions of years, releasing heat that can help keep the center hot. Other processes, including the crystallization of rocks, release heat and help warm up the center. Venus, next to the Earth, but closer to the Sun, has a surface that is a constantly changing sea of lava. In fact it is believed that most of the surface of Venus is less than 250 million years old; the surface has been re-worked by volcanic processes in this time, being melted and laid down again in lava plains and volcanic mountains.

Mars has also had an incredible history of volcanism. Today the surface is pockmarked with volcanoes on an immense scale. Olympus Mons, 21,000 meters high, is the highest mountain in the Solar System, two and a half times the height of Mount Everest. The climbing challenge isn't that formidable though, because the angle of much of the surface is less than six degrees, although at the base of the volcano are impressive scarp cliffs several kilometers high. In other words reaching the summit of Olympus Mons would be more like a gentle hill walk than classical Everest-type mountaineering challenge, once you got over the cliffs. The size of these volcanoes is believed to result from the fact that Mars has had no recent plate tectonics and so hot-spots of magma sit under the same location for great lengths of time, continually spewing lava onto the surface of the planet and causing a build up of these immense structures. On Earth, because plates are moving, the volcanoes can never get the chance to build up in one location for long enough to create structures of this size. Regardless of their imposing sizes and their interest to potential future mountaineers, these Martian volcanoes bear testament that during the early history of our Solar System volcanoes were rife

MORE ABOUT MARTIAN MOUNTAINEERING

The volcanoes on Mars are so high that, in the case of Olympus Mons, the top third essentially projects into space. If you could make your way to the summit you would be surrounded not by the salmon-colored sky of Mars, but by the blackness of space. There are others too that have been formed by hot-spot plumes such as Arsia Mons, which is about 17 kilometers high. Most of the very high mountains are localized in a region called the Tharsis region, a giant anomalous bulge in the surface of Mars. Future mountaineers will undoubtedly come to revere the challenges of the Tharsis region. Volcanoes on Mars have more than just exploratory significance though. Future geologists might climb these structures to reach their vast calderas, many of them with sheer cliffs several kilometers deep. They might sample the insides to understand about the evolution of volcanoes on Mars. The study of these huge structures might help us understand how volcanoes operate on Earth.

on its rocky innermost planets. Volcanism has been a basic fact of life.

To understand how a volcano could cause extinction, one has to know something about how they form. When volcanoes were first studied it was thought that they were exclusively associated with the edges of continental plates. Where one plate disappeared under another the less-than-perfect join would cause molten rock to come rushing to the surface. This idea wasn't inaccurate in the least. Precisely this process caused many of the great volcanoes.

The remote Alaskan Aleutian island chain, the tropical islands of Micronesia and the wilderness of the Andes mountains of South America are all part of the Pacific "Ring of Fire". About three-quarters of the world's volcanoes, including Mount Saint Helens that exploded with such dramatic displays in 1980, occur around this ring of tectonically active plates. The Pacific plate that pushes below the Eurasian plate is responsible for the dormant Mount Fuji in Japan and the bubbling hot springs of Russia's Kamchatka Peninsula owe their existence to the imperfect collision of continental plates. Every year about sixty volcanoes erupt somewhere on the Earth and the majority of these are in the Ring of Fire.

In the twentieth century volcanoes began to be mapped better than ever before as tools improved and geologists began to study the record of past volcanic eruptions. An interesting and unexpected observation was made. Volcanoes were not just associated with the spreading apart or crashing together of continental plates. They could be found in the middle of the continents themselves. The reasons for this didn't seem to be very obvious. Volcanoes suddenly erupting in the middle of continents with no obvious irregularity from which the liquid rock could spring?

The source of these volcanoes appears to be plumes of liquid rock or magma that bubble up from the interior of the Earth. If you have ever seen a lava lamp, you'll get some idea of what I am talking about. Blobs of oil rise up through the lava lamp like the heads of mushrooms with long tails behind them. In simplistic terms, magma plumes from the interior of the Earth are thought to be like this. Blobs of liquid rock, with a diameter of up to 800 km across, escape from the molten mantle and begin to rise to the surface of the Earth. Eventually they reach the continents on the surface and they try to push through. The continents buckle and rise over the pressure of the plume and, given time, the plume might break through.

Evidence for the plume idea comes from the chains of volcanoes that are observed in the Pacific Ocean. If plumes exist, we would predict that as continents move around above these stationary features, so the evidence for them, in the form of volcanoes burning through to the surface, would move as well. This is precisely what is observed. The Louiseville Seamounts, Hawaii Seamounts and the Tuamotu Archipelago are all from different hot-spots. These chains of islands, along with some other members of their groups, all describe an L shape across the Pacific that has taken over 70 million years to form, suggesting that the plate on which they reside is moving over several stationary hot-spots underneath.

Unlike asteroid and comet impact events, scientists do not have to rely entirely on speculation to try and understand what large-scale eruptions might do to the environment. Volcanic eruptions on scales that are large enough for us to observe climatic effects have occurred in recorded human history. Eruptions over the last fifty years, which have happened during a time when our measurement techniques (and training of geologists!) have been good, have enabled researchers to study small climatic changes during volcanic eruptions.

Volcanoes can inject dust and ash into the atmosphere and, if they are powerful enough, these plumes of light-blocking material can reach the stratosphere and achieve global circulation. The effects of this ash and dust are similar to the proposed effects of smoke and dust thrown up by an asteroid or comet impact. A volcanic "winter" might be caused when the reduction of sunlight would cause the Earth to cool.

The so-called "winter" is not necessarily extreme. Recent eruptions of large volcanoes have only caused small temperature drops. Mount Saint Helens (Plate X) apparently caused an average temperature drop of one-tenth of a degree around the world, and Mount Pinatubo perhaps three-tenths of a degree. You can see that these drops in temperature are small enough to make these measurements quite controversial.

As one goes back further in history it is possible to find records of more profound changes in climate. In April 1815, the Indonesian volcano, Tambora, erupted, throwing one hundred and fifty cubic kilometers of ash into the atmosphere, the most violent and the largest ash volcano that has been properly documented in human history. For up to 600 kilometers from the volcano darkness lasted for two days. The ash was thrown so far into the atmosphere, about 50 kilometers, that it reached global circulation. For months afterwards people across Europe reported the incredible sunsets and twilights caused by the dust in the atmosphere. Sunspots could be viewed with the naked eye. The year after the eruption became known in many locations around the world as "the year without a summer". Temperatures dropped and a widespread failure of crops occurred, causing famine and unrest. Across North America records from organizations like the Hudson Bay Company show 1816 as the coldest year recorded. Tambora had given the world a glimpse into the aftermath of a volcanic winter.

Further back in history the records become less reliable, but in AD 536 a writer in Mesopotamia observed that "the sun was dark and its darkness lasted for eighteen months; each day it shone for about four hours, and still this light was only a feeble shadow ... the fruits did not ripen and the wine tasted like sour grapes". The mystery cloud that appeared that year from the Mediterranean caused anomalously cold weather that was recorded as far away as China. In Italy senator Cassiodorus wrote, "The sun ... appears of a bluish light. We marvel ... to feel the mighty vigor of the sun's heat wasted into feebleness. We have had ... a spring without mildness and a summer without heat". There is a spike in acidity in Greenland ice cores from about that date that may well correspond to these events. The spike would have been caused by a large eruption of ash. The volcano, Rabaul, near Papua New Guinea, has been proposed as a culprit.

A common feature of these historical writings is the pronounced drop in light and temperatures that occupy the minds of the writers. This is perhaps not surprising because the effects of these changes on agriculture would have been profound and this is the most easily observed and important environmental change for humans.

Even these eruptions, however, have been relatively benign compared to what the Earth is capable of delivering. Most of the volcanoes that you are I are familiar with, like Mount Saint Helens and Pinatubo, have been of the island arc variety, caused by the imperfect joining of continental plates. This is also true of the more violent volcanoes I've just described from past times. Hot-spot volcanoes caused by plumes breaking through to the surface of the Earth have been predicted to cause massive eruptions of ash and lava which contain high concentrations of sulfur. This element can contribute a great deal to cooling the Earth by forming sulfuric acid droplets high in the atmosphere that help to block out the light. These volcanoes may make the events of Tambora or the year 536 pale into insignificance.

It won't surprise you to know that one can measure how big a volcanic eruption is with a scale, similar to the more familiar Beaufort scale for wind strength. The volcanic explosivity index (VEI) is the scale we use. A VEI of 0, for instance, would be the Hawaiian Kilauea volcano, which has a plume throwing lava about one hundred meters into the air and eruptions on a day-to-day basis. A VEI of 5 is a more serious affair with a plume about 25 kilometers into the air and about one cubic kilometer of lava being thrown out. The explosivity index reaches its maximum at VEI 8, a volcano capable of ejecting thousands of cubic kilometers of ash and lava. Eruptions with VEI 8 only occur rarely, perhaps every few million years and it is these that concern us here.

Sixteen million years ago, a fissure was caused by the eruption of a plume of magma to the surface of the Earth near what is today the Columbia River in North America. From the one hundred kilometer gash in the crust of the Earth emanated one hundred cubic kilometers of lava per day. It could have lasted several days and possibly months. It is known as the Roza Flow and surely had dramatic effects on climate, perhaps cooling the Earth by up to four or five degrees centigrade by injecting large quantities of sulfur-bearing dust and ash into the high atmosphere. The hot-spot has moved about 200 kilometers since it was formed those sixteen million years ago. Today it sits under Yellowstone National Park, USA, providing the heat that drives the hot springs, geysers and other geothermal features of the Park that draw in tourists from around the world. Few of the people who visit Yellowstone will appreciate the violent beginnings from which this quite deceptive beauty arose.

Once temperature reductions on the scale of a few degrees centigrade occur we move into the realms of possible catastrophic effects on the biosphere and the Roza Flow is by no means the largest that has been suggested. Across the surface of the Earth there is geological evidence for flows much larger than the Roza. These "Traps" as they are called (after a Nordic word meaning stair steps, because many of the lava flows today are eroded and resemble large staircases), may have released as much as ten times the quantity of lava thrown out by the Roza Flow. Traps are believed to be regions where hot-spots of magma first made it to the surface. The effects on climate would have been dramatic. Computer simulations have shown that the ash plumes generated by these volcanoes could have reached the stratosphere, thus achieving global circulation. Temperatures could drop by ten degrees centigrade or more. The temperature reductions are not the only problem. It is suggested that the quantity of ash thrown into the atmosphere could be so great that light levels would drop enough to stop photosynthesis. The biosphere is denied its primary energy supply and the knock-on effects would be rather formidable.

So if such eruptions have occurred in the past and they can cause such huge environmental effects, then why don't we find these Traps associated with mass extinctions? And this is where the controversy begins, because indeed, as we saw in Chapter 4, we do find Traps associated with extinctions in the fossils. The Deccan Traps, a million square kilometers of basalt lava that cover an area of western India from Nagpur to Bombay on the western coast, up to two kilometers thick in some places, coincide with the K/T extinctions 65 million years ago. The Siberian Traps, in the northern central part of Russia, coincide with the end-Permian extinction, which killed 95% of all species on Earth 250 million years ago. The Siberian Traps are as impressive as the Deccan Traps. Covering an area of just less than half a million square kilometers they are more than three kilometers thick in some places.

The remarkable results found by scientists around the world who were examining these Traps were how quickly they seem to have been formed. The Siberian Traps may have been laid down in less than one million years and the Deccan Traps in less than five million years and possibly in less than half a million years. Such eruptions in such small time scales must have had dramatic consequences for the environment and hence the interest that surrounds the fact that some of the volcanic Traps seem to coincide with periods of great biological change.

A volcanic winter might not have drastic effects on all of life on Earth. Imagine if a volcanic winter occurred during the middle of natural winter when many deciduous trees have lost their leaves and gone into a state of hibernation anyway. If the volcanic winter occurred at the beginning of the winter and lasted for six months, then it is quite possible the effects on many plants could be negligible. They would emerge from the winter not knowing that anything untoward had ever happened. But for bigger volcanoes the light-blocking aerosols and particles are predicted to stay around for possibly a year or longer, particularly for some of the large Traps that may have erupted many cubic kilometers of lava a day over many years and possibly thousands of years. The severe drop in average temperatures together with prolonged reductions in light would be likely to affect even the hardiest plants on land and the microbes in the sea. So although volcanoes are less instantaneously catastrophic than asteroid and comet impact events, their effects may be more damaging over prolonged periods and perhaps just as capable of bringing catastrophe to the biosphere.

During the formation of the great Traps these challenges to the biosphere may have been repeated many times over the course of a million years or less as eruptions would have been large and episodic over a geologically short period of time. It is imaginable that some species, whose numbers could have dwindled during these changes in the environment, could be pushed to extinction. If the first period of darkness and reductions in temperature didn't kill them off, then a second period of the same environmental assaults might do the job.

The sulfur thrown up in the eruption might form sulfuric acid in the high atmosphere and rain down. In the shallow parts of the ocean and in lakes and streams some groups of microbes, such as the Foraminifera with their shells of calcium carbonate and their sensitivity to acid conditions, might be killed. Because carbon dioxide is a greenhouse gas and large quantities of it are expected to come from the erupting volcanoes, global warming might ensue after the volcanic winter. The destruction of the ozone layer by chlorine compounds from the magma might plausibly cause an ozone hole and increase ultraviolet radiation at the surface of the Earth. Added together, these assaults might have been too much for many species, and have driven them to extinction.

You can see that I've used the word "might" five times in the above paragraph! The computer models suggest that all the above are possible and maybe even probable. Scientists have some idea of the compositions of lava and magma and they can calculate the quantities of gases that could be ejected by large Trap eruptions. So, despite the lack of observations of giant eruptions, there is good reason to suspect that in the past some volcanic eruptions have been big enough to be able to cause extinctions.

As I have mentioned once or twice before, you are probably beginning to see that many of the effects expected from a large Traps eruption are quite similar to asteroid and comet impacts. The long winter, the acid rain and the ozone depletion may be effects common to both catastrophes. Can we ever tell them apart in the fossil record? Separating the environmental effects of both calamities might be a very difficult thing to do, particularly at extinction boundaries where people have proposed that there might have been an impact at the same time as volcanism (like the K/T boundary). Although the environmental effects might be quite difficult to separate, separating the physical evidence for volcanoes and impacts is probably easier. Impacts leave signs that are not seen with volcanoes. Shocked rocks, formed by the sudden collision of an asteroid or comet with the Earth at high speed are quite specific to impacts. Volcanic eruptions can cause fires in their vicinity, but they are not believed to light global wildfires in the same way as impact events are supposed to be able to do.

The involvement of impacts and volcanoes in extinction is still controversial, mainly because of this problem of demonstrating a direct causative link between the catastrophe, the environmental changes and the extinctions. At the K/T boundary there seems to be good evidence for an impact occurring and for the Deccan Traps as well. There is even some geological suggestion in India that the asteroid impact occurred during the time the Deccan Traps were being laid down. An iridium-rich layer has been found between two episodes of lava outpouring from the Traps. Maybe the impact was another nail in the coffin for the biosphere, an unfortunate coincidence that the Earth was on a collision course with a large asteroid just when vast volcanoes were causing changes in the biosphere.

A more extraordinary claim has been made that impact events are the cause of large-scale eruptions. The premise behind this idea is that asteroid impacts would either punch a hole in the crust, causing an eruption, or that the shock wave from the impact would travel around the Earth and focus on the opposite side, causing a massive eruption. So far there is no evidence for this idea. The Chicxulub impact crater, the supposed crater of the asteroid that contributed to the K/T extinctions, is not exactly opposite the Deccan Traps. Although it is true to say that some of the great lava plains on the Moon, for example, are associated with the sites of giant asteroid and comet impacts, these seem to be special cases of particularly massive impact events during the early history of the Solar System when the bodies of the Solar System were hotter and more molten than they are today. At the moment it doesn't seem likely that even an asteroid of ten kilometers in diameter, which might be the size of the one that caused the Chicxulub crater, could cause a giant volcanic eruption.

The importance of volcanoes as agents of extinction during our journey around the Galaxy can be better appreciated by just looking at the statistics. Over the last 225 million years, the period of our journey once around the Galaxy, there have been up to nine major Trap episodes of volcanism. Compare this to the expectation that during this same time period the Earth would be exposed to one close supernova explosion and just two large-scale impact events on a scale that may cause global extinctions. Indeed one could argue that our journey around the Galaxy is a journey dominated by the challenge of surviving the Earth, interspersed with insults from the cosmos. This is why I chose to include volcanoes in this book, because without them the story is skewed. You would be left with a vision of life on Earth at the mercy of the cosmos alone, but the true picture also requires an understanding of the Earth. I don't feel guilty about this concession because, as I explained earlier, volcanoes are a result of the way in which the Earth formed. The evolution of the Galaxy and the Solar System is very much a part of why life on Earth is faced with the problem of volcanism and thus, in the broadest sense, the challenges it poses to the evolution of life is a legacy of our galactic origins.

Microbes as a whole would be little affected by the onset of periods of great volcanism. If the effects I have described are real, then the worst to come off are photosynthetic microbes that require light to produce energy, but the subsurface biosphere will surely be hardly affected at all. The shut down of light, the reductions in temperatures, acid rain, ozone depletion and the eruption of liquid basalt over millions of square kilometers, these events must be catastrophic for the surface biosphere, but for a microbe several kilometers in the crust feeding off hydrogen and carbon dioxide? Volcanism will not affect these microbial communities, even in its most intense manifestation.

The microbial world is even more robust than this. Not only did it almost certainly survive in the subsurface unscathed from the most violent volcanic episodes, but it no doubt took advantage of the volcanic Traps as well. Volcanoes create habitats for life and maybe you are now beginning to see a parallel story with that of the chapter on asteroid and comet impacts, where we saw the microbes that can colonize impact craters.

Yellowstone National Park is today the surface manifestation of a hot-spot of magma beneath the Earth's crust (Figure 18). Within it are geothermal springs that range from cool streams to boiling cauldrons up to 95 °C. The Park is a playground for tourists who come from around the world to admire this unspoilt region of the United States of America and to see for themselves an extreme environment. Many imagine that this is how the early Earth might have looked during the early Archean, 3.5 billion years ago.

FIGURE 18. Volcanoes destroy life, but they also offer new opportunities. Hot springs in geothermal regions like Yellowstone National Park (USA) give us an insight into how the microbial world can take advantage of the habitats created by these events. (Image, Axel Rosenberg.)

Within all of these hot pools are thriving microorganisms. Octopus Spring is a particularly interesting pool. The spring is a riot of color. At its source, where the temperature is 95 °C, are filaments of a pink microbe that clutch the rocks in the stream and twist and turn in the emerging water. Genetic investigations show that it is primitive. It groups with other extremely primitive microorganisms of the archaea group. Further from the outflow, as temperatures cool to about 55 °C, cyanobacteria colonize the pond. These heat-loving or "thermophilic" microbes grow best at these hot temperatures and form luxuriant mats of green *Synechococcus* that spread out across the pond. Flies and spiders daintily run around on the surface of the mats avoiding the hot water beneath them.

The story is repeated consistently in the geothermal springs of Yellowstone. Recent work by Norman Pace, Sue Barnes and others in their groups have shown that these springs have a vast microbial diversity. One might expect that in such extreme volcanic environments diversity would be low because only a few microbes could eke out a living, but to many microbes adapted to life at high temperatures these are not extreme environments but, in fact, the norm. The diversity is instead very high. Many hundreds of species of microbes might inhabit a single pond. At different temperatures different groups come into play and dominate the pools.

In one stream, which is extremely acidic with a pH less than 2, like vinegar, and a temperature of 55 °C, lives an alga called *Cyanidium*. It has an intense green color and the stream is a bright green hue, as if someone had played a practical joke and poured fluorescent-green food coloring into the stream. In other locations in the Park are giant white mounds of calcium carbonate and silicon compounds laid down by thousands of generations of microbes. These terraces of "travertine", made by the microbes, are testament to the domination of these hot thermal environments by the microbial world.

MORE ABOUT MICROBIAL FOSSILS

Unsurprisingly, the fossilized remains of microbial communities that live in volcanic or "hydrothermal" regions (that would also be found around the edges of fresh smouldering asteroid or comet craters) have caught the interest of those seeking life on other planets. Microbes are so small that when they die, they leave very little trace. The pressures and heat that the remains of microbes are subjected to as the rocks they reside in are subjected to geologic processes make the preservation of their remains unlikely. However the mounds they create from the compounds they excrete or from the compounds that crystallize around them when they are growing in hot springs provide a sign of their presence. These features are large enough to be seen with the naked eye or quite easily under microscopes. Maybe we could look for these fossil features on Mars? Trying to understand the way in which microbes fossilize is an important area of investigation for many people, including petroleum geologists who want to understand how fuels are formed from the remains of dead microbes.

Yellowstone is not exceptional. In the hot springs of Rotarua in New Zealand, similar microbial mats of *Chlorobium*, a microbe that photosynthesizes using sulfide from the spring water, can be found. The microbe grows in streams at about 55 °C, taking advantage of the sulfide that emanates from the geothermal sources. On the island of Lutao near Taiwan in the Ring of Fire, microbial communities live in springs ranging from 50 to 90 °C. Some of these microbes are not only tolerant of high temperatures, but live in outflows with two percent salt, meaning they are also very salt tolerant.

Volcanic regions are host to more than just streams. In the caldera of volcanoes, lakes can form that last for thousands of years. Crater lakes have been studied all over the world and the findings yield a kaleidoscope of microbial life. Depending on the influences of the chemistry of the lake, they can harbor great diversities of microbial communities. Crater Lake in Oregon (Figure 19) was formed just over

FIGURE 19. Crater Lake in Oregon, USA, is home to a lake that sits in the caldera of a volcano, just one of thousands of volcanic crater lakes around the world. Warmed by the geothermal heat from the volcano, the lake is home to a rich diversity of microbial life. (Image, Axel Rosenberg.)

6,500 years ago, although there were small eruptions as recently as 4,000 years ago. The six hundred meter deep lake is home to rich microbial mats that metabolize the iron produced from hydrothermal fluids that circulate into the lake from the volcanic regions beneath. The microbes form long filaments that intertwine to produce the mat structures that hold them to the bottom of the lake. Similar microbial mats occupy the thermal springs around the Loihi Seamount at the far eastern part of the Hawaiian chain of volcanoes. The hydrothermal fluids are a source of energy for these communities. Because heat drives circulation through the lakes, they are provided with a continuous conveyor belt of new food.

Some animals can take advantage of volcanic eruptions too, like the northern pocket gophers that invaded the cooling volcanic ash of Mount Saint Helens a few years after the eruption. But they are confined to the period when the region has cooled down. The highest temperature at which a multi-cellular animal has been recorded is about 55 °C. As the microbial world has its upper temperature limit at 113 °C, it can invade the smouldering pools and springs of volcanic regions almost as soon as they are formed. Under these circumstances it is impossible to imagine that volcanoes could make the microbial world extinct, because to many microbes the eruption of a fissure of molten lava heralds the beginning of a new habitat; a welcome escape of the cold world at temperatures less than 60 °C and the opening of a new and bountiful existence at higher temperatures.

8   Intelligent stupidity

There is a threat to the biosphere that is perhaps as great as any that the Galaxy can contrive. The rise of intelligence on a planet brings with it the inevitable consequence that the intelligent organism will use its intellect to manipulate resources, to find ways to use the resources to expand its population. And the manipulation of resources almost invariably involves modification of the environment. Humans are not a cosmic threat, but as a product of the last half a million years of the galactic journey, we should talk about humans and their ability to rival the threats from the Galaxy.

Humans have been supremely successful at modifying their environment. Only during the last three decades have we begun to come to the terms with the idea that six billion badly behaved apes might be capable of modifying the atmospheric composition of the planet itself, usually with negative consequences. Pollutants such as chlorofluorohydrocarbons (CFCs) deplete the ozone layer and increase the penetration of ultraviolet radiation to the surface of the planet. Carbon dioxide, the gas generated from burning oil, coal and other organic matter, increases the temperature of the planet by acting as a greenhouse gas, letting heat in, but slowing its escape to space. In 1998 humans released about 6 billion tonnes of carbon dioxide into the atmosphere.

The fact that humans might be able to alter the atmosphere is actually hardly surprising. The atmosphere is a thin veneer that hugs the surface of our planet. Just a few tens of kilometers high, the height of the bulk of the atmosphere is about quarter of the distance from San Francisco to Los Angeles or the distance from London to Oxford. Seen in these terms it is perhaps not so difficult to see that the injection of millions of tons of pollution into the atmosphere every day might eventually be able to modify this tiny coating of gas on our world.

In our quest to get access to resources (particularly metals, oil and gas) and in our quest to build towns and cities we modify the rest of the planet as well. Forests are felled and lakes are drained to make space for fields and buildings. The wholescale modification of habitats around the world has led to the extinction of species. It is estimated that about 30,000 species go extinct every year, most of this being caused by the destruction of biologically diverse forests in tropical regions. In reality it is difficult to be certain about this figure because habitats are being destroyed before they have been studied, so scientists just don't know how many species many habitats have. These extinctions are believed to be so profound that some have classified the period we are now living in as a period of mass extinction. Some write of it as the next great extinction of the Phanerozoic eon; the "sixth extinction", following the five that have brought devastation to life on Earth in the past, some of which we met in Chapter 4.

If there are something like 50 million species on Earth (this may well be a gross underestimate) then it will take a mere 1200 years to achieve something like the destruction of 95% of species, which would put our current era in a league with the great end-Permian extinction 250 million years ago. Extinctions are usually associated with the decline of species over hundreds of thousands or millions of years, but humans may be capable of doing this over much shorter time spans. And the rate of generation of new species is not believed to be increasing to make up for any of this loss. All of these calculations are highly controversial and it is not the purpose of this book to delve into the environmental consequences of human behavior on extinction rates. Nevertheless, you can see that the rates of extinction that mankind can bring to bear upon the biosphere over hundreds or thousands of years put us in the league of agents that are easily capable of delivering a mass extinction.

Our impact on the rates of extinction, particularly of large animals, has been an emerging problem for tens of thousands of years.

This is short in geological time, but human society was threatening species long before there were six billion of us and a proliferation of global industry. Eleven thousand years ago two-thirds of all large mammals, including sabre-tooth tigers and mammoths, went extinct in North America in just a few thousand years. The causes might be a multiplicity of factors, including some changes in climate, but the discovery of arrowheads in remains of mammoths and computer simulations of human intervention in animal populations suggest that the arrival of humans in the Americas may have contributed significantly to the demise of species during the Pleistocene. These large, easily captured animals would have been obvious prey for the new arrivals.

The introduction of domesticated species into habitats has been a common problem with human expansion and is often a cause of the demise of animals at the hands of human society. The warrah, a fox-like animal that lived in dens in the Falkland Islands in the South Atlantic, became extinct in 1876 after the introduction of sheep onto the islands. The Island government offered bounties for killing the foxes to prevent them from menacing the newly introduced sheep. All that remain of this rather wonderful-looking beast is two stuffed exhibits in natural history museums in Belgium and Sweden. Cats have been claimed as the reason for the extinction of the dodo in Mauritius. The twenty-kilogram birds, evolved from African fruit-eating pigeons, went extinct around the beginning of the eighteenth century. Originally thought to have been driven to extinction by human hunting and the culinary excesses of Portuguese sailors, it now looks as if cats, which would have easily hunted these giant flightless birds and found their eggs, drove them to the edges of the island and ultimately to extinction.

Human beings themselves now join impact events and massive volcanism as one of the great threats to large surface-dwelling animals. But the humble microbe, as we have seen throughout this book, is not so easily threatened by the behavior of humans. Like impact events and volcanism, humans may threaten the large animals, but much of our destruction and waste is a new opportunity to the microbial world.

Each one of us in the Western world produces an astonishing half a tonne of household waste every year. The USA has to deal with about 230 million tonnes of waste each year when all the refuse from businesses is added into the final equation. Just less than half of this waste is paper, followed by food and general garden waste and then there are plastics and metals. All of this waste has to go somewhere and in most countries it gets buried. About half of all the waste produced goes into landfills, the rest being incinerated and hopefully some (but less than a third while this book was being written) is recycled.

So that's still an astonishing 100 million tonnes of waste that needs to get buried in the USA alone and for the whole human population we can only guess, but it is on the order of billions of tonnes of waste each year.

The waste we produce gets collected and poured into landfill sites. These giant holes in the ground are filled with a layer of clay to help stop poisonous liquid finding its way back into the water supply. A plastic liner is usually put inside to help contain this liquid. Then the waste is piled in, with some intermediate layers of soil until the landfill is complete.

The landfill (Figure 20) is like a cake to many microbes and even small invertebrates like flies. Imagine a hole with layers of food and paper and then the odd layer of soil thrown in for good measure. This underground gateau, almost as soon as it if finished, is a home to vast microbial populations.

The first microbes to arrive are fungi and other microbes that feed on food and paper. The paper-eating microbes are essentially the same types that eat dead trees, digesting the cellulose inside tree trunks. Inside the landfill they find a plentiful supply of ground-up trees in the form of magazines, newspapers, chip wrappers, candy covers and boxes. About 900 million trees are felled each year to supply the US need for paper and about a third of these end up as food for microbes inside landfills.

After about eighteen months the landfill becomes depleted of oxygen as it is used up by the microbes and animals. The microbes

FIGURE 20. Landfill sites are the result of the hundreds of millions of tonnes of waste and refuse that build-up each year from human activity and within them microbial populations thrive. They are one of the fastest growing microbial ecosystems created by man.

that need oxygen to survive begin to die off, but as they do so a whole new collection of microbes adapted to survival in low-oxygen conditions begins to move in and continue eating the landfill. The broken-down products begin to be converted to alcohols and various acids, a veritable concoction of nasty liquids. The liquid so produced is called leachate. It concerns us because it can leach out of the landfill into the groundwater and potentially contaminate our drinking water. This is why landfills have the clay and plastic liners.

As the process goes on so the acids and alcohols are converted to acetate; each waste product in the microbial chain is food for the next microbes as the products that were originally magazines, plastics and kitchen waste get broken down into smaller constituents and consumed. The breakdown process releases nitrogen and phosphorus that help the microbes grow even faster. Eventually the acetate becomes food for methanogens, microbes that turn the compound into the gas methane, which can build up in the landfill. Now, as you might be aware, methane is quite explosive, so one environmentally sensible way to deal with this problem is to extract the methane from the landfill and use it to generate power. This new solution is now being used more often to deal with the landfill gas problem. Methane is also a greenhouse gas, so letting it simply escape into the atmosphere is not a good idea. Like carbon dioxide, which is also produced by landfills, it can contribute to global warming.

The microbes will go on producing the explosive methane for about 10 to 60 years until they run out of food. It is not known with any great certainty what happens in the very long term because our consuming lifestyles are a relatively recent phenomenon, but once the bulk of the landfill has decomposed, it will settle down. It can even be used for building on again. JFK airport in New York is built on an old landfill site.

You might be thinking to yourself that perhaps this isn't such a bad environmental problem after all. If microbes like to eat the waste, can't we use them to our advantage to help us deal with the environmental problems we face? Indeed this line of thinking has become a field of intense interest and activity. It's called "bioremediation" and the use of microbes to help us degrade toxic compounds is a new alliance between the microbial world and humans.

The involvement of microbes in landfills isn't always to our benefit, though. Our attempts to subjugate the microbial world to our own uses will always be fraught with the complication that some of them are unlikely to simply bow down and cooperate with us. Research workers at Oakridge National Laboratory in the USA found that microbes could use methylate mercury compounds, turning them into highly toxic methyl mercury. This compound can be released as a gas into the environment at concentrations about 30 to 40 times higher than it would normally be in air. Methyl mercury causes birth defects and although the raw elemental mercury from which it comes is also very poisonous, mercury tends to stay in the landfill sites more effectively than the methylated variety. Microbes can take advantage of our waste as food, but the waste they produce in the process can, in some circumstances, threaten us even more.

Microbes have come into a win–win situation. Either we will join them as allies and purposefully feed them our waste or we will cover ourselves in a mound of waste until we ourselves go extinct and the waste will still be available to microbes living naturally in the environment. The story of the landfill is a remarkable one; an insight into how even an intelligent species cannot bring devastation to the microbial world. Waste produced by us might poison some ecosystems and some microbes may well suffer locally from our indiscriminate dumping, but this is more than counterbalanced by the opportunities we have created. By digging, purifying and manufacturing resources from deep within the earth, we have, if you like, accelerated the availability of many minerals and compounds to microbes that might otherwise have been short of these resources.

Most microbes are eventually limited by something in their environment. This is good news because if they weren't we would soon be overrun by a mass of microbes. Humans can disrupt this balance by making a nutrient available that previously was limiting their growth. Once this happens, microbes will begin to grow in earnest until they are limited by something else.

Three-quarters of the air we breathe is nitrogen; a rather unreactive gas, but one that is essential for a whole diversity of biological processes. Nitrogen is incorporated into amino acids that make up our proteins and so it is essential for growth. The problem is that most organisms cannot use the nitrogen directly as a gas from the atmosphere; they need it to be "fixed" into more readily used sources, like nitrates and ammonium compounds, found in the food they eat. There are a group of microbes called nitrogen-fixing microbes that can transform the atmospheric nitrogen into available nitrogen compounds and every year they turn about 140 million tonnes of atmospheric nitrogen into nitrogen compounds that are then available to the rest of the biosphere, including us.

A further five million tonnes of nitrogen compounds comes from lightning strikes. Lightning has the energy to transform nitrogen in the atmosphere into oxidized nitrogen compounds that then rain out of the atmosphere and become available to the biosphere but, as you can see, its contribution is small compared to what microbes are responsible for. Nitrogen fixation is one of the most important processes on our planet and it appears to have been around for a long time. One group of microbes that can fix nitrogen are the cyanobacteria that are at least three billion years old. Found in many locations around the world, one most important association is with rice, the staple food for much of Asia. By fixing the nitrogen from the atmosphere they make nitrogen available to the rice, which can then grow. This clever association is found in the roots of many vegetable plants like beans that have a symbiosis with nitrogen-fixing bacteria in their roots; the bacteria provide available nitrogen and in return the roots provide nutrients for the microbes.

If this process went on unchecked, the nitrogen from the atmosphere would eventually be consumed and so a feedback process operates, driven by another set of microbes that convert nitrogen compounds from dead organisms back into nitrogen gas.

And so, through fixation and denitrification, a subtle balance in the world's nitrogen budget is maintained. The effect of human activity on this nitrogen cycle has been quite dramatic. Humans now add about another 150 million tonnes of available nitrogen into the biosphere each year – as much as the natural cycle. It is added as fertilizers and some of it comes from power plant emissions. Once added to fields, fertilizers rarely stay where they were put, eventually leaching into groundwater and out into rivers and streams where they can make their way into the oceans, seas and lakes. In the North Atlantic Ocean basin, about 40% of the supplied nitrogen comes from humans.

Human-produced nitrogen can cause real problems in places that previously had very little of this nutrient, particularly in seas like the Baltic and Black Sea that are enclosed. The newly added nitrogen can trigger algal blooms, which are a thick soup of algae. The algae can release toxins, killing off other marine animals, or they can reduce the sunlight getting through the water, killing off plant life on the sea or lake bottom.

The reason why some algae can produce toxins is not really known with certainty. One species of algae, *Pseudo-nitzschia*, produces the toxin domoic acid, which has no known function as yet. Speculation focuses on its role as a defense mechanism or a way of getting rid of excess energy or even as a mechanism to bind trace metals that it might need for growth. Domoic acid is toxic to other microbes and fish and the spread of this organism caused by nitrogen dumping could increase the number of toxic blooms that occur in our water systems. The focus in this book is not human pollution, but the survival of the microbial world. Although I speak of toxic blooms and pollution, the very word "bloom" will tell you that something is thriving on our nitrogen waste. The tiny toxin-producing flagellate *Gymnodinium* that blooms in the wake of a nitrogen release is doing well for itself, even if at the expense of fish, other microbes and human health. Like the landfill, there is rarely a situation that is detrimental to all microbes. What is an environmental disaster to many organisms is likely to be an environmental utopia for many species of microbes.

We often talk of ecosystems being destroyed by human activity but, as should now be apparent, this is quite an anthropocentric view of the world. In many cases we have actually created new and luxuriant ecosystems for other organisms. Many microbes, if they could express appreciation, would probably laud the vast expansion of the landfill "ecosystem", one of the fastest growing ecosystems on Earth. And there are other ecosystems on the increase as well.

Consider roads and parking lots. The construction of new highways can be a matter of public outrage. Because they often involve the felling of trees and the plowing up of countryside they are useful symbols to the environmental community of the destruction we impose upon the rest of the biosphere in the name of progress. However, the laying of asphalt offers new opportunities to microbes that consume hydrocarbons, the oils from which asphalt is made. The asphalt ecosystem is one of the fastest growing ecosystems on Earth today. Not since the oil reservoirs of the world were laid down in the depths of the oceans by the sacrificial bodies of marine microbes, have microbes that use hydrocarbons seen such a great expansion of their available habitat.

Before the oil and asphalt reaches the roads to form this great new ecosystem for microbes it has to be transported as oil around the world and unfortunately we are not perfect at this. In 1989, the oil tanker *Exxon Valdez* crashed into a reef and released millions of gallons of crude oil into the Prince William Sound in Alaska. Oil that was, until then, safely trapped in deep sediments was suddenly released into the oceans and contaminated 5,000 kilometers of shoreline. Between 100,000 and 300,000 birds were killed and some fish populations were halved. The eleven million gallon spill was one of the greatest environmental oil catastrophes that has been seen. Many techniques were used to clean up the oil. It was burned, it was manually washed off animals and rocks and, finally, it was fed to microbes. Microbes in nature that degrade hydrocarbons will, given enough time, degrade an oil spill. Taking the microbes to the oil directly can accelerate the process. After the spill, the oil was sprayed with the fertilizer EAP 22, which contains nitrogen and phosphorus. The nitrogen and phosphorus provide the natural oil-consuming microbes with the nutrients they need to get to work breaking down the substance. Very soon the microbes were cleaning up the mess.

Of all the waste products humans produce, perhaps the most unpleasant is our nuclear waste. And it's important to understand that it's not just nuclear power plants and weapons that contribute towards this stockpile. X-ray machines from hospitals and a variety of other useful machines produce low-level nuclear waste that slowly piles up. Most of this low-level material is buried or compacted because it is not too toxic and a lot of it decays quite quickly, so in that respect it is certainly less of a problem than the waste from nuclear reactors and weapons.

High-level waste that forms in the cores of nuclear reactors is mainly made up of the element uranium. The 450 nuclear power plants around the world produce about 12,000 tonnes of high-level waste each year. The waste includes plutonium, cesium and strontium produced from the fission decay of uranium in the reactors. And these products can last for a long time. The half-life (the time for half of the radioactivity to go) is about 30 years for the strontium, but for the plutonium it is about 24,000 years! This is why we have to find ways to bury this waste for long enough for the radioactivity to reduce to safe levels, a challenge that has occupied many organizations testing concrete, synthetic rock, glass and other methods to encase the nuclear waste before it is buried.

It won't surprise you by now to know that some microbes can deal with many of the waste products from nuclear processes and have been investigated as a possible way to clean them up. Microbes found in soils can reduce uranium into forms that are more easily removed from groundwater. Many of these microbes are the sulfate-reducing microbes that we met in the deep-sea vents in Chapter 3. Given some nutrients, the natural sulfate-reducing microbes like *Desulfovibrio* will begin to get to work on the uranium. Radioactive elements from radium to thorium can also be removed from soils by these microbes. Others, like *Citrobacter*, can remove plutonium from water as well. All over the world there are a variety of microbes that have the ability to transform one or more radioactive elements. So far there does not seem to be one element produced by our nuclear industries that can't be used by at least one microbe somewhere.

The waste products of our predilection for war offer a veritable menu for microbes. During the 1990s it was found that

TNT (trinitrotoluene), used in a range of bombs, can be degraded by microbes. Over 90% of TNT from soil samples could be removed in about three weeks by providing the microbes with some sugars and phosphate to help them grow.

Nuclear reactors and munitions produce quite an extreme form of waste, but many other materials are also dumped into the environment. Zinc and copper are just two everyday metals that are toxic at high concentrations and there are many places in the world where copper extraction has led to very high concentrations of the element in the environment, that prevent normal plants and animals from surviving.

In Montana, USA, there is a lake near Butte (Figure 21). A placid lake, it is the sort of spot that looks idyllic to someone who knows little of what the water contains. In 1995, two hundred snow geese flying over the lake spied its placid waters and, tired of their long

FIGURE 21. Berkeley Pit near Butte, Montana, USA, is an acid pit full of cadmium, zinc, cyanide, copper and a concoction of chemicals that, overnight in 1995, killed two hundred geese that landed on it. Yet the lake is home to a diversity of microbes, including a variety of algae, that can tolerate the chemicals and high metal concentrations. (Image, Robert Ashworth.)

journey during bad weather, landed on the lake. The next morning all 200 geese were found dead on the lake; they had drunk the pit water filled with copper, arsenic and cadmium. Berkeley Pit, as it is called, is the site of an old copper mine that, in the late nineteenth century, provided a sixth of the world's copper supplies, a region of unusually pure and easily accessible copper. It was mined then and continued to be mined until well after World War Two, although the depths needed to get even moderate-quality ore became deeper and deeper. In the 1950s the deep-mining stopped and the methods switched to extracting copper from the soil with water and cyanide solutions. The mine was shut down in 1982. The sulfide compounds in the ore were, until they were mined and exposed on the surface, protected from the oxygen in the atmosphere. Mining exposed the compounds and they turned to sulfuric acid, helped along by microbes in the soil. The now acid, cyanide-containing waters from the mines in the area are filling the two-and-a-half-kilometer-wide Berkeley Pit at a rate of about three million gallons a day and the local council now has a race to find a way to reverse the pollution before the pit overflows.

Today the lake is very acidic. It has a pH of 2.5, almost as acid as lemon juice. As well as the cyanide compounds it contains cadmium, zinc, copper and aluminum and a variety of other metals. However, the lake is not a dead lake. Within it are algae, including species of *Euglena* and *Chlorella*. Adapted to living at high metal concentrations, the algae have mechanisms to either exclude the metals from within themselves or to collect them at high concentrations and immobilize them to prevent them from causing damage. By "complexing" the metals with metal-binding compounds that they excrete or produce inside the cells, the metals are no longer freely available to move around in the cell and cause damage. These microbes that can make a living in the waters might soon be artificially grown in laboratories and put to work cleaning Berkeley Pit and reversing the environmental damage.

The ability of these microbes to live in such toxic environments is often because they have plasmids; small circular pieces of

DNA that contain a cassette of genes that allow them to process toxic compounds. A microbe called *Ralstonia metallidurans*, which was first found in industrial settling tanks in Belgium in the 1970s, has two plasmids that encode genes allowing it to detoxify zinc, cadmium, cobalt, lead, copper, mercury, nickel and chromium, just the sort of genes you might need for a microbe to go into the Berkeley Pit. Because plasmids are small pieces of DNA they can be passed from one microbe to another by sexual combination allowing the resistance to move to new microbes. In many ways this is exactly how antibiotic resistance works. Plasmids containing resistance to antibiotics can be transmitted between strains of microbes. This is why there is such a problem with antibiotics in hospitals, with the growing number of microbes that are becoming resistant to the drugs. Microbes can develop resistance to many different types of antibiotics, meaning that a whole suite of antibiotics can sometimes become completely ineffective. An "arms race" develops between medical science and the microbial world as humans try to keep up with the developing resistance to new antibiotics as it spreads throughout the microbial world.

As with antibiotic resistance, *R. metallidurans* perhaps acquired the ability to survive in metal-poisoned environments by acquiring the DNA that has encoded on it the necessary pathways from some other microbe. Even the ability to degrade oil is carried out by a set of genes on a plasmid. Plasmid pHCL can degrade hydrocarbons and could be useful for dealing with oil spills. One group of scientists has managed to transfer the plasmid from a microbe called *Pseudomonas* into marine microbes to produce a series of microbes that can be used to degrade oil spills on the oceans. The beauty of plasmids is that they are more easily isolated than genes that might be buried in different places in an entire genetic sequence of a microbe. The circular pieces of DNA can be readily tracked down and that makes it easier for us to use them to produce designer microbes to clean up our waste. The flip-side is, of course, that these plasmids are more readily transferred from one microbe to the next and partly explain the ability of the microbial world to adapt so rapidly to the pollution that we generate.

The direct manipulation of microbes to do our cleaning up for us has become very interesting to the nuclear industry. A team in the USA has managed to take the genes from a strain of microbe normally found in our gut and incorporate them into the extremely radiation resistant *Deinococcus radiodurans*. The genes from the gut microbial strain encode a process that detoxifies mercury ions that are produced during nuclear weapons manufacture. The newly engineered microbe can take the mercury ions and turn them into elemental mercury, which is more easily disposed of. Because it has the radiation resistance of *D. radiodurans* it can survive some of the most intense radiation environments that nuclear-processing facilities have to offer.

Ironically, it is humans that are now contributing to the impossible extinction of microbes. By engineering microbes that can tolerate and even thrive in some of our most terrible wastes, we are not only giving ourselves a chance to clean up, but ensuring that the microbial world remains robust against our worst insults on the environment.

The microbe *D. radiodurans* (called by its fans D-rad for short) is sometimes, crassly, but deservedly, called "Conan the bacterium" in honor of its toughness. It was originally found by Arthur Anderson at Oregon Agricultural Experimental Station in Portland, USA in 1956. Anderson first noticed and then isolated the red-colored microbe from a can of ground meat that had become spoiled even though it had been sterilized with megarads of radiation. The microbe can survive a radiation dose about 3,000 times greater than the dose that would kill a human. As well as in cans of irradiated meat, the radiation-hardy microbe has been found growing in radioactive cobalt-60 storage tanks in Denmark. How does the microbe survive such incredible radiation doses?

D-rad's ability to survive seems to lie in the ability to literally string together the pieces of DNA that were broken during the radiation damage. Most cells have the ability to repair DNA damage if one strand is broken. If both strands are broken, then the cell can die. A typical microbe cannot tolerate more than five double-stranded breaks. However, D-rad can deal with up to 120 double-stranded breaks in its DNA molecule and still survive. Within about twelve hours following exposure to radiation, the fragments of DNA are restored. It manages to do this with a very clever set of repair processes that can put the DNA back together again. It can compare the broken strands of DNA with other copies of DNA, which are most likely to be broken at a different place. Once the split strands are matched to other copies, the broken pieces are put back together.

The microbe can prevent the fragments of DNA from wandering off into the cell before they are properly put together. It appears that it has a type of molecular scaffolding on which the pieces of DNA can be assembled as they are put together again, thus allowing the microbe to re-assemble its genetic information, a bit like the scaffolding that might be used to repair the roof on your house.

How did D-rad end up with such remarkable DNA repair abilities? Some speculate that it is because the microbe is also found in some extreme desert regions of the world. Its ability to survive damage in reactor storage ponds came about because of the need to repair damage caused by extreme desiccation every time it dried out. There is nowhere in nature, they say, that these organisms would have been exposed to intense nuclear radioactivity at the sustained levels that would have caused these repair abilities to evolve.

Although the desiccation theory is quite compelling, one needs to be careful about our rather egocentric view. We sometimes almost take pride in the fact that we have created environments that microbes have never seen before. But the waste products from our nuclear society may not be absolutely new to the microbial world.

A remarkable discovery was made in 1972 in the African country of Gabon. Deep within a uranium mine, scientists found evidence for natural nuclear reactors (Plate XI). Gabon is home to rich seams of uranium ore laid down when the Earth formed 4.5 billion years ago. The Oklo uranium ores were found by geologists in the 1960s and they are located in the south-east of the country in the grassy highlands of this small African state. In some places in the uranium beds, the concentrations are so high that they are up to two-thirds uranium oxide. Two billion years ago, when the uranium had not decayed as much as it has today, it was giving off enough neutrons to start nuclear fission. Fifteen fossil nuclear reactors have been found in the mines now and some of them such as reactor number 9 have been studied quite thoroughly.

It was while studying the rocks from Gabon that a French analyst working at the mines found that they were very slightly depleted in uranium-235, suggesting that some of it could have been lost in nuclear fission reactions. Further investigations showed that

MORE ABOUT FOSSIL NUCLEAR REACTORS

The way the fossil reactors in Gabon worked is as follows. Uranium-235 is a compound that can cause a chain reaction. If it absorbs a small subatomic particle (that rain down all the time from space) it can split into what are called fission products and they give off two or three more neutrons. These two or three neutrons then collide with other uranium-235 atoms and cause them to split. A chain reaction is initiated that leads to nuclear fission, the basis of energy generation in nuclear reactors and nuclear weapons. Uranium comes in two forms. One form, uranium-235 is much less abundant than the heavier form, uranium-238. In fact today only 720 uranium atoms in every 100,000 are uranium-235. But uranium-235 is the most fissionable material and it's the form required to get nuclear fission going in reactors. Today, to make nuclear reactors work, the uranium has to be purified to enrich its uranium-235 content to be useful for nuclear fission. Two billion years ago the Earth's uranium deposits were much richer in uranium-235 than they are today because less of it had decayed (in fact scientists use the decay of uranium to date rocks). During that time the uranium beds in Gabon had deposits that were rich enough in uranium-235 to cause nuclear fission, hence nuclear reactors were operating naturally on the early Earth.

the rocks also contained large quantities of fission waste products such as plutonium-239. The presence of neodymium, an element produced during fission, at exactly the same concentrations as those expected of nuclear reactors, is one of the most compelling pieces of evidence of nuclear fission. The water that circulated around the deep subsurface rocks at the time acted as a nuclear moderator, reflecting the neutrons into the reactor core and enhancing the fission reactions. These reactors may have operated for as long as half a million years.

Even more extraordinarily, the existence of plutonium showed that the reactor had even bred its own fuel, the plutonium fission product itself being highly fissionable. Thus, the Gabon reactors were in essence equivalent to modern breeder reactors.

Just above the reactors are layers of quartz that formed from the hot water that circulated around the layers of rocks when the fossil reactors were operational. Could they have contained microbes that evolved to survive in nuclear reactors? Did some microbes trapped in fluid inclusions near these reactors evolve DNA repair processes to deal with survival under intense radiation conditions? All these questions are speculative and I don't wish to assert without any evidence that radiation-resistant microbes might have emerged from natural nuclear reactors. The point I want to make is that the versatility of the microbial world is more than enough to deal with the waste products of intelligent species. Even some products that we might consider the most man-made and unnatural products of our advanced society such as nuclear weapons waste, have been seen by nature before.

It is unlikely that humans can contrive to make the microbial world extinct. Every wrong turn we make, every error in environmental judgment we impose upon our world is a new opportunity for some species of microbe.

9   The world is not enough?

We can, if we try, find scenarios that will make microbes extinct. The end of the Earth, consumed in the fireball of the Sun as it turns into a red giant, will make all microbes on Earth extinct in about five billion years time, if nothing else succeeds in the meantime. You might be wondering why it is, then, that the title of this book is *Impossible Extinction*. The reason why we might say that the extinction of microbes is impossible is because of the long-standing debate that microbes might already have escaped the Earth altogether. Carried on rocks ejected by impact events, is it possible that these vehicles transfer microbes between planets in our Solar System, and perhaps leave the Solar System altogether? Of the places in our Solar System that hold the most promise for answering part of this scientific question, Mars is undoubtedly the most compelling. Did life transfer to Mars from Earth and vice versa?

The question of life on Mars is still speculative, but even the possibility of life on another planetary surface in our Solar System is important for our discussion here. The significance of the search for life on Mars to our understanding of extinction and the fate of biology during the galactic journey is that life on Mars would provide robustness against a number of events that threaten life on Earth. We have already seen that many microbes are, for the most part, essentially invulnerable to extinction events on Earth. Apart from a massive ocean-sterilizing impact that was able to cook the Earth down to six kilometers, thus extinguishing the deep-subsurface biosphere, there is no event that can remove microbial life from Earth. The effects of supernova explosions are confined to the surface of the Earth and the effects of volcanism are also essentially confined to the surface, with little influence on the deep subsurface. However, the presence of a biosphere on Mars would certainly provide robustness against even the unlikely event of a massive microbe-extinguishing impact event on Earth. It would require near-simultaneous impact events on Earth and Mars to extinguish microbial life altogether.

Mars captured the imagination long before scientists had the means to survey it for life. The red color of the planet, caused by the rusted surface of iron oxides, made it an unusual sight in the night sky and the Romans and Greeks, associating this color with blood, named it in honor of the God of War, Mars (or Ares in Greek). The detection of lines on the surface of the planet would later turn out to be an optical illusion. During the nineteenth century it was not appreciated that when two objects close together are viewed from far away, they will tend to look as if they are joined by a line; an optical processing trick of the human brain. These lines, connecting different desert regions of Mars, so voluminously written about by Percival Lowell as "canals", fired the imagination of the nineteenth-century public. They were happy to believe that they were the works of a desperate civilization seeking to tap the last remaining vestiges of water and channel it to their beleaguered cities. H.G. Wells further embodied this view of the Martian surface in his book, *War of the Worlds*, a tale of an attempt by this desperate Martian civilization to invade and conquer Earth. The book's interest is not only sociological, probably reflecting the fear we have of invasion, but it was the first time the public had truly been aroused into a state of discussion and excitement about the possibility of life on another world. Although its scientific basis was non-existent and it spawned speculation about life on Mars that even today continues to impair serious scientific attempts to address this question, it did much to bring the question of extraterrestrial life into the public mind.

By the beginning of the space age, the vagaries of telescopic observations could be supplemented by cameras orbiting the planets themselves. The first robotic spacecraft to visit Mars in the 1960s returned images of a dry and lifeless world. No canals, no civilizations and no large bodies of liquid water. But ironically, the low resolution of the Mariner 9 images had pushed our perceptions of this barren world to the other extreme. What was missed in these images would be revealed in the Viking images in the 1970s and more so in the Mars Global Surveyor images of the 1990s.

Mars is indeed a barren world. There is no obvious sign of life on the planet. The atmosphere is apparently, at least in its gross characteristics, not out of equilibrium with chemical processes. On Earth, the oxygen in the atmosphere is testament to a gross imbalance caused by the production of oxygen by photosynthetic microbes. The presence of ozone in the atmosphere, made possible by the high oxygen concentration, is yet another signature of the widespread activity of life. An alien civilization looking for signs of life on Earth could seek out the ozone signature by examining the spectrum of light given off by our atmosphere that would contain the signs of ozone – a clue that life was at work on our planet. On Mars, no such imbalances are detected. The atmosphere is 95% carbon dioxide and there is a mere 0.13% oxygen in the atmosphere. But lack of imbalance in the atmospheric composition does not provide evidence for lack of life, merely that if it does exist, the extent of its biological cycling of gases is much lower than that on the Earth and to a point where it is extremely difficult to detect.

The images returned during the 1970s began to show intriguing signs of a world that might not be so quiescent as expected in

MORE ABOUT LOOKING FOR SIGNS OF LIFE IN THE ATMOSPHERE

Some microbes that live in the deep subsurface of the Earth use hydrogen as an energy source and get their carbon from carbon dioxide. These "methanogens" produce the gas methane as a byproduct (we met the same microbes in Chapter 8 living in landfill sites). One way to search for life in the deep subsurface of Mars might be to use spacecraft with sensors highly sensitive to methane and see if the gas can be detected coming out of cracks in the Martian surface. Wisps of water vapor could also be sought in subsurface gas emissions.

the 1960s. The Viking orbiters provided pictures of what looked like channels carved out across the Martian surface by a liquid. Long since dried out, many of these channels appeared to have regions at the end of them where the frequency of craters was less than in the ancient plains around them, suggesting that the craters had since been somehow removed. The craters had been smoothed over by a liquid, suggesting that perhaps liquid water had flowed across the planet at some stage in its past. Were these lakes at the end of ancient giant rivers? Teardrop-shaped formations around some craters suggested that at some time in the past they had sat in the wake of rivers or catastrophic floods, which raged around them.

Tiny channels at the end of larger channels looked like deltas. On closer examination it began to become apparent that many of the features one associates with rivers and streams were represented in one form or another on the dried plains of Mars. A new epoch opened in the exploration of Mars as a new question began to emerge. If Mars is lifeless now, did it perhaps have life in the past, when liquid water was more abundant? Do we have an example of a planet where life has gone completely extinct? This question brought with it the requirement that early Mars was warmer than it is today. With an average temperature of below freezing across most of the surface, such rivers and lakes would be impossible to maintain today and even at the equator where temperatures during a balmy summer's day peak above freezing, the atmospheric pressure is too low for large bodies of water to form. Perhaps Mars had a high concentration of carbon dioxide in its early atmosphere that could act as a greenhouse blanket, keeping the temperatures above freezing and pushing the surface pressures high enough for large bodies of liquid water to be stable. These discussions sparked the fierce debate amongst planetary scientists that continues today.

From a biological perspective the prevailing view emerged that Mars was a planet that once could have supported life; at least it may have met the theoretical requirements for life, even if ultimately it proved to have remained sterile. Our knowledge of the conditions on early Earth seemed to fit the idea quite well. After all, early Earth was supposed to have had an atmosphere that was mainly carbon dioxide with very little oxygen and it had abundant water. And we know that there was life from the abundant evidence in the fossil record. It might be reasonable to suspect that Mars, with conditions similar to early Earth, could once have supported life.

Higher-resolution pictures of the channels and valleys would be needed to look at this question more successfully and in the 1990s Mars Global Surveyor returned remarkable images of the Martian surface. Objects as small as a meter across could be detected. Better pictures provided further credence to the idea that liquid water once flowed across the Martian surface. More stunning discoveries were recent gullies, seepage around the edges of impact craters that bore testament to the possible presence of a liquid seeping out from the sub-surface of Mars, even today (see Figure 22).

The gullies on Mars re-open the question of life on present-day Mars because they suggest the availability of liquid water, which, with an energy supply and the necessary elements, is required for life. Energy appears to be in abundant supply on Mars. There is sunlight and even in the absence of photosynthesis, the different iron compounds in the soils can provide a source of elements for chemosynthesis. No doubt many other elements and compounds such as sulfates could be used as well. Hydrogen and carbon dioxide may be present in the Martian subsurface and these could also supply the raw material for energy acquisition by a subsurface biosphere, just as they do in the deep subsurface of the Earth today.

The rest of the basic elemental building blocks for life seem to be on Mars. The Viking Landers examined the soils of Mars by scooping up samples and subjecting them to X-ray fluorescence spec-trometry. The elements, iron, silicon, sulfur, magnesium, potassium and many others that are required to build biological systems were found. Nitrogen has remained rather elusive, but it may be hidden away under the dusty surface. Nitrogen constitutes 2.7% of the Mar-tian atmosphere, so we know the resource is there in gaseous form. Based on what we know of the chemical requirements for life on Earth, Mars does seem able to support life.

FIGURE 22. The Yogi rock at the Mars Pathfinder site is typical of the photographs that suggest that Mars is a barren desert planet. But there are more tantalizing glimpses of an active world. Bottom left shows a valley network image taken by the Viking Orbiter in the late 1970s. The networks reveal the past presence of water. Today there is evidence even for near-surface liquid water in the form of recent gullies around the edges of impact craters, such as these in Noachis Terra at 30° S (lower right). (Image, NASA.)

Although the chemistry looks good, on the face of it the physical conditions on the surface of the planet don't look as appealing. Because of the lack of oxygen in the atmosphere there is virtually no ozone, the gas that protects life on Earth from most of the harmful ultraviolet radiation that would otherwise reach the surface of the planet. A small amount builds up over the Martian poles, but apart from that, there is little. The damage caused by ultraviolet radiation to a piece of the genetic material, DNA, on the surface of Mars would be about a thousand times as great as it is on the Earth.

Powerful oxidants that destroy organic molecules may be rife in the Martian soils. One of the great mysteries of the Viking Lander experiments in the 1970s is why they didn't find evidence of organic molecules. One wouldn't necessarily expect life, but scientists know that they should expect some organic material, because carbon-rich meteorites constantly bombard the planets. Over billions of years there should be abundant evidence for them on Mars. But there isn't any evidence for organic molecules and, in fact, they seem to have actually disappeared. Part of the explanation may lie in their destruction by the ultraviolet radiation, but to explain the Viking Lander biology experiments, many hypothesize that the soil is full of powerful oxidants that will react with organic molecules. The presence of these oxidants would help to explain why there are no organic molecules on Mars today, but they do not bode well for the chances of surface life.

It is thought that these challenges are really quite ubiquitous across the surface of the planet. The Martian surface is covered in a fine layer of dust, created from the erosion of rocks over the 4.5 billion year history of the planet. Giant dust storms that can rage for a third of a year and cover the entire surface of the planet ravage Mars and intermittently local dust storms can cover an area the size of North America. The oxidant-laden dust is transported across the planet and deposited over the surface as a uniform veneer.

But another, more formidable, challenge to life across much of the surface is the lack of liquid water. On much of the surface of Mars liquid water cannot exist. This is because the pressure on the surface of Mars is so low. When water-ice is heated at pressures below 6.1 millibars (about a hundredth of Earth's atmospheric pressure) it turns instantly into vapor, bypassing the liquid phase altogether. The average atmospheric pressure on Mars is almost exactly 6 millibars and so in most locations, even if there is ice, there would still be no liquid water should temperatures became sufficient to melt it.

In many respects these extreme physical conditions are in good agreement with what is observed. One would predict from our knowledge of terrestrial microbiology and the environmental conditions on Mars that life would have a hard time existing on the surface and this agrees with what we see. Bizarre life forms are not found there. We don't find any atmospheric disequilibria suggesting gross atmospheric alterations caused by abundant surface life. We find a dead surface.

But are there habitats where life could exist? Are there habitats where microbes could still be hiding and making a living, escaping the extinction-causing conditions on the surface of Mars?

The extremes of the surface of Mars are not too difficult to get away from. Just a few millimeters under the soil the ultraviolet radiation is reduced to insignificant levels. The depth the oxidants go down into the soil is not known for certain. Some speculate that they may get several meters into the Martian subsurface, but some speculate that it is just the top few millimeters in the dust layer. Nevertheless, at a depth of just a few meters and probably a lot less, it is likely that the oxidant problem is gone and the soils become a less stressed environment for any potential microbe. The liquid water is the most intriguing question. On the surface there is plenty of evidence that the planet is an icy world. Polygon shapes in the soil, that are also seen in the Arctic and are caused by successive freeze–thaw in the soil, cover some of the plains of Mars. Mounds, called pingos, have been tentatively identified. These are features of frozen polar land and are formed by the collection of water into the center of a mound that grows into a ice lens, pushing the ground up from underneath. All of these features suggest that frozen water is in the subsurface. Although planetary scientists don't know whether there is abundant liquid water beneath the surface of Mars, they have good reason to believe there might be, as the seeps and gullies in the craters testify.

The best places to look for life on Mars would therefore be in the subsurface. As we saw in Chapter 3, the subsurface of Earth is teeming with microbial life. Drill cores at depths down to 5.3 kilometers have shown the presence of microbes. The only limiting factor seems to be the high temperature. Once it reaches about 110 °C, it becomes too hot for life to exist. As Mars is smaller and cooled down more quickly than the Earth, it is possible that the conditions for life might exist much deeper in the Martian subsurface.

You could look for this life by drilling into the subsurface much as people do on Earth. Drill rigs could be assembled by robotic craft. The problem with this approach is that drills are very complex pieces of equipment and need constant maintenance. The difficulty of drilling even several hundred meters into the Martian surface might be too great for a robot to accomplish. Deep drilling might have to wait until humans arrive on the Red Planet.

But another way to look in the subsurface is to use nature's drill – impact craters. There are some quite nice equations that relate the diameter of a crater to the depth at which material is excavated to the surface within easy reach of robot or human explorers. You can calculate the size of craters you would need to prospect to particular depths in the Martian subsurface to look for signs of life. A crater with a diameter of 260 kilometers would excavate material down to a depth of at least five kilometers, corresponding to the deepest regions of the terrestrial subsurface biosphere. There are some candidate craters on Mars that fit this criterion. Newton crater, at 41° S, in the heavily cratered southern hemisphere of Mars, has a diameter of 287 kilometers and material surrounding this crater might tell us something about whether the deep subsurface of Mars supported a biosphere.

A crater inside a crater is a good way to look for life. DaVinci crater near the equator is 107 kilometers in diameter but within it is a much smaller crater. If DaVinci had ever contained a lake conducive to life, the smaller crater in the deposits could have excavated material from the subsurface of these deposits and brought it to the surface. Perhaps these deposits contain within them evidence for sediments from an ancient lake?

If there is found to be no life on Mars, the scientific implications are still really quite interesting. The public are often led to believe that scientists are going to Mars to seek life and if they don't find it they will be very disappointed. Scientists seek answers to questions and a good scientist is not too concerned with what the answer is, they merely wish to find the truth. But human beings aren't this perfect and finding life on Mars would undoubtedly cause some great excitement. A positive answer will open up whole new vistas of research into the nature of the life on Mars. So one can understand why scientists might prefer to find life on Mars.

Proving there is no life on Mars is actually rather a difficult thing to do. If you search in a variety of habitats and find no life you might think there is no life on Mars, but in reality you would have to scour every square centimeter of the planet to definitively say that there was no life. But say scientists looked in the most obvious places to find evidence of past or present life; the deep subsurface, ancient crater lake beds, ancient deltas and hydrothermal areas near volcanoes, say we looked in all these regions and found no evidence of past life. What then? For the galactic journey the result might be significant, because it would suggest that life neither evolved nor was successfully inoculated on Mars. Catastrophic-scale extinction events on Earth that destroyed microbial life would be the end of the line for microbes. There would be no second chance on Mars.

The search for life on Mars is a biologically important quest. It can tell us something more than merely whether life exists on another planet. It gives us an insight into how life fits into the cosmic picture of evolution. Its ability to survive catastrophes of large magnitudes and particularly massive impact events, is dependent upon its presence on multiple planets. We get a better idea of what exactly it would take to make life completely extinct.

Before I end our discussion, I must talk briefly about some other worlds where people have speculated life is present. One of these is of particular interest because it is known for certain that it is now a lifeless world, but it may once have been more conducive to life.

Venus (Figure 23) is a boiling cauldron. Today, the temperature of the second planet from the Sun, a constant 464 °C across its surface, is

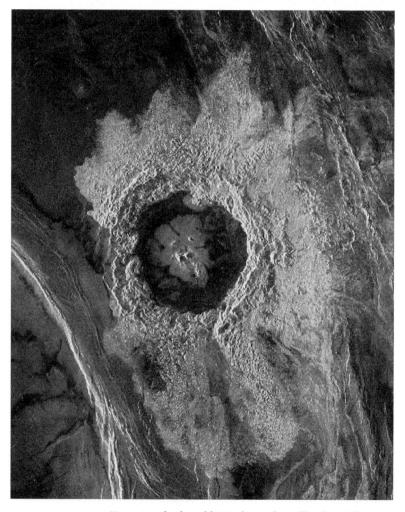

FIGURE 23. Venus is a dead world. No chance for gullies here. The complex crater, Dickinson, like the rest of the surface of Venus, is devoid of water and baking at a temperature of 464 °C, about 350 °C higher than the presently known upper temperature limit for life. (Image, NASA.)

well above the upper temperature limit for life. Even if mechanisms did exist to allow biochemical reactions, the surface has no liquid water.

Like the atmosphere of the early Earth and Mars, Venus has an atmosphere of carbon dioxide, but the difference between Venus and these other two planets is that it is much closer to the Sun. Its orbital distance from the Sun is 108 million kilometers compared to the Earth's 150 million kilometers and Mars, which has a mean distance of 228 million kilometers. This close to the Sun, the high concentrations of carbon dioxide on Venus act as an effective blanket, creating a runaway greenhouse affect on the planet so that the temperatures rise above the upper limit for life. The pressures on the surface are also quite extreme, equivalent to 100 atmospheres (but the pressures aren't in themselves enough to prevent life, because many microbes are quite happy at these sorts of pressures. The Mariana Trench in the Atlantic, home to many pressure-loving microbes, has a pressure ten times higher than this).

Venus may not have always been this way. In the early history of our Solar System the Sun was about a quarter less luminous than it is today and the runaway greenhouse effect may have been much less intense on Venus. Some scientists have suggested that Venus may have been home to a moist greenhouse, where the temperatures were close to boiling, but nevertheless the planet was covered in oceans like the early Earth. The low ratio of hydrogen to its heavy counterpart deuterium has been taken as evidence that Venus once had a huge inventory of water, which it subsequently lost. The water would have been split up in the intense greenhouse as the Sun got hotter. The hydrogen in the water was lost to space and the oxygen recombined with the rocks.

A profound question is posed by this view of a clement early Venus – was it a planet where microbial life became extinct? Is a runaway greenhouse effect the one mechanism that can even penetrate to subsurface life and cause a planetary-scale extinction event of all microbial life? When the Sun gets hotter, a global extinction of all microbial life will occur on Earth, in about five billion years time. This is the

MORE ABOUT LIFE ON VENUS

The study of Venus has had a colorful past. At the turn of the twentieth century it was thought that the planet might have harbored swamps ripe for microbial life. Later, when it was realized that the planet was likely to be very hot, researchers suggested that the planet was covered in boiling oceans and that near the poles, acid-loving, heat-tolerant microbes might live in the oceans. Even after it was discovered that the surface was dead, speculation did not cease. Carl Sagan suggested that organisms like float bladders containing hydrogen could float around in the atmosphere of the planet. For one brief evening, after some good drinking with two of his friends, he founded the Society for Venusian Biologists, which the next morning, in the cold light of day, ceased to exist. Venus is a dead world, but its chemistry is still a fascinating benchmark for thinking about what extremes life can tolerate and what it can't.

fate of our own biosphere and of that we are certain. Venus's history is a window to our biological future and the inevitability of extinction.

The problem with life on Venus is that we are unlikely to ever prove that extinction did occur on the planet. Venus is a geologically active planet. Its molten interior has re-worked the surface and today most, if not all, of the surface is less than 250 million years old. Eruptions of lava cover its surface. Any remnants of early land masses that might shed light on the existence of early microbial life, using chemical techniques, similar to those geologists can use on early rocks of Earth, are likely to have been destroyed. Perhaps, like our finding of a 3.5 billion year old Martian rock, the famous ALH84001 meteorite that was thought to contain Martian life, an ancient Venusian meteorite might one day be found. Until ancient Venusian rocks are found, the idea that Venus was a planet that experienced a global scale microbial extinction will remain within the realms of speculation.

Optimism is, I suppose, an admirable human trait, and one that has definitely had a pervasive influence in astrobiology, the study

MORE ABOUT ASTROBIOLOGY

Astrobiology is a rapidly expanding field that essentially covers "life in the Universe". It is a subject that covers many of the links between life and the cosmos, from the origins of life through to the effects of asteroid and comet impacts on the Earth. The beauty of the subject is that it provides an environment for interdisciplinary thinking. In some ways it is like the classical courses in natural sciences in its breadth of subjects, but it focuses on contemporary questions at the interface between biology and planetary sciences. The word "astrobiology" was first used in 1953 by Gavriil Tikhov, a Russian scientist, to describe the study of life on other planets, but its remit has greatly expanded and it is now taught in many universities around the world.

of life in the Universe, which has more recently expanded to cover many other aspects of the links between life and the cosmos. Although the surface is dead, that hasn't stopped people from wondering about life in the atmosphere, floating around in the cloud cover of Venus. As one rises into the Venusian atmosphere the temperatures drop and become more conducive to life. Between altitudes of 48 and 57 kilometers the temperatures lie between 0 and 60 °C, quite pleasant for microbial growth. The likelihood that there is life in the clouds of Venus is very low, though. Although there is some water in these regions, it is tightly bound into sulfuric acid droplets that have a concentration up to 98%. It would be very difficult for a microbe to extract water in sufficient abundance from these droplets and the sulfuric acid would be very damaging to organic matter. Organisms that could use sulfate and hydrogen to make a living and that like oxygen-free conditions would stand the best chance of making a living in the Venusian clouds, but even for them, the conditions on Venus are probably too extreme.

The only other possible refuge for life that is known about in our Solar System is on a moon. The ice-covered moon Europa, a satellite only slightly smaller than our own Moon, orbits the giant gas planet Jupiter. Observations of this moon, not only from the surface

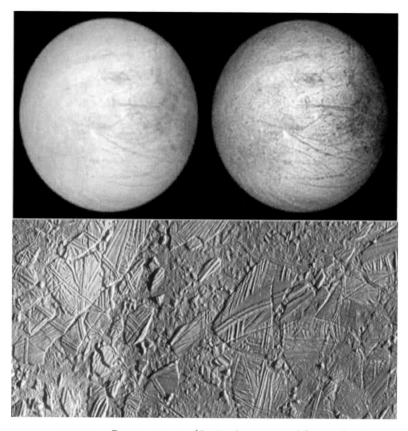

FIGURE 24. Europa, a moon of Jupiter (upper image), has a subsurface ocean. Rafts of ice (lower image) break up and re-freeze on the dynamic surface of this satellite. These oceans are an important focus for the search for life. (Image, NASA.)

of the Earth, but also from the Voyager spacecraft in 1977 and the Galileo spacecraft in 1999 suggest that it has a crust of water ice (Figure 24). The constant buckling of the moon by the tidal forces of Jupiter not only generate heat to melt this ice, creating an ocean under the crust, but it also makes this ocean a potential abode for life. Life could dwell in the oceans, feeding off organic material delivered by impact events or formed on the surface of the moon by ultraviolet radiation, which then leach down the multitude of cracks in the icy surface. The warmth of the oceans could provide opportunities for organisms on the ocean floor using hydrothermal vents to get their elemental needs, in an analogous way to the communities of the hydrothermal vents of the deep-ocean floor of the Earth. Life in the deep oceans would not have enough light for photosynthesis, but these other nutritional modes could be supported if one assumes that Europa has its inventory of the typical elements found in other planets and moons.

Probably the only way to search for life on Europa is to land on its surface. Perhaps one could land near a crack in the ice and sample the deep-ocean water as it washes up and down inside the crack during the tidal movements of the moon. Alternatively a submersible that would melt its way into the ice and through to the ocean could look for life, uncoiling a communication line as it goes. The vehicle would carry out tests on the water, perhaps taking photographs of filtered water samples in a miniature microscope or testing for biological molecules in the water. The technologies now being developed to study the deep subglacial Lake Vostok in the Antarctic, which I talked about in Chapter 3, could perhaps be applied to studying the oceans of Europa.

Of course, even if Mars and Europa did have life, we don't know whether it would be related to life on Earth. If it arose from a completely independent origin then talk of second chances on other planets becomes irrelevant because one would be dealing with a different origin of life and therefore completely different biospheres. But what if microbes had been transferred between these other planetary bodies and the Earth? During the last decade there has been a great deal of interest in the possibility that life might have been transferred from planet to planet by asteroid impacts.

To begin an interplanetary or interstellar journey, a microbe first needs to find itself in a situation where it is inside a rock that will be ejected into space. As we have already seen, the subsurface of the Earth is thriving with microbes living in the spaces inside rocks. Because microbial life is quite ubiquitous down to one kilometer depth and has been found as deep as five kilometers, almost all rocks that are excavated during an impact event will contain some microbes.

During an asteroid or comet impact the ground is heated to enormous temperatures, up to 15,000 degrees centigrade, and subjected to great pressures, but although one might expect all the microbes involved in an impact to be killed, not all the rocks thrown out of the crater are heated and shocked right the way through. Like a badly cooked pie, there are pockets in the material that is thrown out that remain at low enough temperatures for microbes to survive.

The badly cooked pie anology leads to the notion that microbes could be ejected at low enough pressures and temperatures to survive a journey off the surface of the Earth, a journey fast enough to escape the Earth's gravitational field altogether. Spores of the microbe, *Bacillus subtilis*, fired from a rifle into a clay target show that the acceleration to reach the speeds needed to escape the Earth's gravitational field (at least eleven kilometers a second) isn't any barrier to some microbes being launched on a journey into space. In some of these experiments nearly all of them survived.

The journey itself isn't insurmountable. Provided the spores are shielded from ultraviolet radiation, which is the most damaging radiation to their DNA during the long interplanetary flight, then they can probably survive long enough for a few to make it to Mars. In experiments *Bacillus subtilis* spores managed to survive the vacuum and cold of Earth orbit for six years. If they are inside a rock or covered in just a thin layer of dust or meteoritic material, then they have a good chance of making it across the interplanetary distances or surviving like this for millions of years if they need to (Figure 25). If this is indeed the case, then haven't microbes been transferred back and forth between the planets all the time? The amount of material that lands on Earth each year from Mars is about 500 kilograms. There will be less from Earth landing on Mars because the rocks must escape the Sun's gravity to travel outwards to Mars, but nonetheless there will be some transfer and it will not be too dissimilar to the amount that reaches Earth from Mars. Perhaps about 10% of this material is launched at low enough temperatures to allow life to survive. And so there is no doubt that there is plenty of material being transferred

FIGURE 25. Can life be transferred between hospitable abodes for life in our Solar System? Martian meteorites such as this one (EETA79001) that have been found on Earth suggest that microbes could be transported through the Solar System. (Image, NASA.)

back and forth each year, which could potentially harbor life and, in the case of rocks from Earth, certainly does.

If there is life on Mars, maybe it's just our cousins then. Of course if life on Mars was completely independent to life on Earth then scientifically it might be more interesting to a microbiologist. However, if it was our cousins, then there is a microbial insurance policy on Mars; a subset of Earth's descendents will ensure escape from extinction on Earth.

There is one extinction that is certain; the extinction caused when the Sun turns into a red giant in five billion years time. At this point the insurance policy comes to an end, if it exists of course. The Solar System itself will face a final perilous existence as Earth evaporates and Mars probably becomes too hot for life. At this point being on other planets becomes unimportant. An impossible

FIGURE 26. A glimpse of our far biological future. The star Betelgeuse, 425 light years away, is a red giant in the constellation Orion (the star is in the top left of the constellation). This is the ultimate fate of our own Sun in approximately five to seven billion years time when the Earth, and all life on it, will be extinguished. But the possibility that microbial life has already escaped the Solar System and become immune to this eventual destiny continues to intrigue many. (Image, Space Telescope Institute, NASA.)

extinction would require making it to another Solar System altogether (Figure 26).

It won't surprise you to learn that there have been some thoughts about the likelihood of life being transferred between Solar Systems. In fact the idea is at least a century old. Svante Arrhenius (1859–1927), a Nobel Prize winner, suggested in 1908 that microbes could be transported within the Solar System and beyond by radiation pressure from the Sun. Without invoking meteorites to do the job he believed that microbes might float out of the atmosphere and get caught up in the pressure of particle streams and light emanating from the Sun and thus be transported through space. His ideas are not fundamentally different from those being suggested today, although most mechanisms behind this process of "panspermia", as it is called, involve the launching of life into space in impact ejecta from asteroid or comets.

You can calculate the number of rocks that get ejected from the Earth by impact events over time and you can then calculate the numbers that leave the Solar System. Then, knowing the distribution of stars in the Galaxy, you can work out the chances that one of these rocks will firstly be intercepted by another star and secondly that it will be propelled into the inner reaches of this new solar system to be captured by an Earth-like world. And, as you can imagine, the calculations are open to huge arguments. How many Earth-like planets are there out there and what constitutes a planet ripe for inoculation by a rock traveling from Earth? Can microbes survive a trip this long and how long is the trip?

The first attempt to look at this problem came to the conclusion that it might be very rare for life to be transported between solar systems. Only one meteorite ejected from a planet in our own Solar System reaches another system every 100 million years. But this hasn't met universal agreement. Some believe that during periods of close encounters with other stars it would be much easier for life to be transferred between solar systems. Like some type of interstellar relay race, stars would pass the baton of life back and forth through their comet clouds every 20 million years or so as they intersected each other. After a time, some of these comets would come onto a collision course with planets in the solar system.

The notion that life might be transported between solar systems is speculative, but it forms the final unknown part of the question of an impossible extinction (Figure 27). Microbes are supremely successful survivors and the physiological attributes they possess undoubtedly mean that they can survive without the Earth if liquid water and the right energy supplies come along. Many display attributes of what we might call "cosmotrophy", the ability to tolerate conditions on other planets or in space. As this is the case, have microbes actually already been ejected and transported to other planets? Have they already, long ago, escaped the unpredictabilities of the isolated Earth, prone as it is to the extinctions that are now so well recorded from the fossil record? Have they even escaped the final,

FIGURE 27. The Whirlpool galaxy, perhaps not dissimilar to the view down onto our own galaxy. Like the Milky Way it contains upwards of 100 billion stars. Has the microbial world achieved an impossible extinction by distributing itself amongst the stars in our own galaxy? (Image, NASA.)

certain, extinction event, the Solar-System-destroying death throes of our Sun in five billion years time? If they have, then the microbial world might lay claim to the accolade of being an impossible extinction.

Numbers and units

100	10^2	one hundred
1,000	10^3	one thousand
10,000	10^4	ten thousand
100,000	10^5	one hundred thousand
1,000,000	10^6	one million
1,000,000,000	10^9	one billion
1,000,000,000,000	10^{12}	one trillion

1 inch = 2.5 centimeters = 25 millimeters
1 foot = 0.3 meters
1 mile = 1.61 kilometers = 1, 610 meters
1 light year = 9.4 trillion kilometers
1 mile per hour = 0.47 meters per second
1 acre = 4, 047 square meters

0 °C =	32 °F
40 °C =	104 °F
100 °C =	212 °F
250 °C =	482 °F

Glossary

acetate   A type of organic salt found particularly in soils.

acid/alkali   A measure of the concentration of hydrogen ions in a liquid. Acid substances have a high concentration, alkaline substances less so. Acids have low pH and alkaline substances high pH.

acritarchs   Loose definition used for any group of microbes that have soft organic walls.

ALVIN   Deep-sea submersible vehicle operated by Woods Hole Oceanographic Institution in the USA. Introduced in 1964.

Archean   The time of Earth history covering the period from about 3.9 billion years ago to 2.5 billion years ago. The period in which the first unequivocal signs of life appear on Earth.

asteroid   A small rocky body orbiting the Sun. Most of them come from the asteroid belt between Mars and Jupiter.

astrobiology   The science of the study of life in the Universe. Covers a broad set of questions that address the links between life and the cosmos.

ATP   Adenosine triphosphate. An energy-rich compound used in organisms for energy storage.

bacteria   Extremely small, unicellular microorganisms that multiply by cell division; occurring in spherical, rod-like, spiral or curving shapes and found in virtually all environments; some types are important agents in the cycles of nitrogen, carbon and other matter, while others cause diseases in humans and animals.

barophile   An organism that needs high pressure to grow.

Biosphere 2   A series of ecosystems enclosed in glass structures north of Tucson, Arizona, USA. Used for experimental studies of ecosystems in controlled conditions.

black hole   A star so dense it gravitationally draws in light, giving it a black appearance.

caldera   The crater in a volcano, usually caused by the collapse of underground lava reservoirs.

carbon dioxide   A colorless gas present in our atmosphere at 0.03%. It is a greenhouse gas, letting heat in and preventing its escape to space.

catastrophism   The viewpoint that the geological or biological history of Earth has been subject to periods of rapid change in a catastrophic manner.

CFC   Chlorofluorohydrocarbons. Produced for refrigerators, aerosols and other appliances and disposable items. These chlorine compounds can break down ozone causing the ozone hole.

chemosynthesis   The use of chemical reactions to get energy.

Chicxulub   A buried asteroid crater in the Yucatan Peninsula, Mexico. The site of the collision associated with the K/T boundary.

Chroococcidiopsis   A type of cyanobacterium that is desiccation and radiation resistant. Found in hot and cold deserts of the world, often living in rocks.

comet   Object made of ice and dust originating in the outer reaches of the Solar System in the Oort cloud.

cosmotrophic   An organism that displays the ability to survive or grow in the environment in space.

Crab Nebula   The remains of a supernova explosion seen in the year 1054. The nebula is 6,500 light years from Earth.

cosmic rays   Highly energized radiation coming from sources in outer space. They consist of protons, electrons and alpha particles and some heavy atomic nuclei.

cyanobacteria   Photosynthetic members of the eubacteria. They are often called blue-green algae, although they are not in the same group as algae.

detritivore   An organism that can eat the remains of other organisms.

dinoflagellates   A group of microbes with two tails that can swim. The group are mixotrophs.

DNA   Deoxyribose nucleic acid. The long molecule that stores the genetic code, the information that contains within it the instructions to build organisms.

electron   A tiny particle in orbit around the nucleus of an atom.

endolith   A microorganism living within rocks, particularly in the subsurface of the rock.

eubacteria   The forms of life generally also called "bacteria". They include many of the common bacteria found around us.

eukaryotes   Organisms whose cells have a membrane-bound nucleus and many specialized structures located within their cell boundary. Includes all multicellular animals, but also some single-celled organisms such as photosynthetic algae.

extinction   When organisms are reduced to numbers where they can no longer be sustained.

extrasolar planet   A planet orbiting another star.

fission   The splitting of an atom into two or more parts.

forams or foraminifera   A group of marine microbes with large shells or tests made from calcium carbonate.

Gal-year   One of the names suggested for the time it takes for the Solar System to go once round the Galaxy.

gene   A piece of the genetic information that codes for a particular protein.

grains (in rocks)   The individual particles that make up a rock, like the grains of sugar that make up a sugar cube.

habitable zone   The region around a star where liquid water is stable.

halophile   An organism that can grow in high salt concentrations.

Hertzsprung–Russell diagram   A diagram that depicts the relationship between the spectral class of a star (its color) and its absolute magnitude – essentially a graph of the life history of stars.

hydrothermal vent   A spring where hot fluids are emitted, often providing a habitat for heat-loving microbes. Either in the deep oceans where continents spread apart or sometimes in asteroid or comet impact craters.

hyperthermophile   An organism that can grow at temperatures greater than 80 °C.

impact winter   The postulated period of dark and cold caused by the injection of smoke and dust into the atmosphere by a large asteroid or comet impact.

inoculate To contaminate something by introducing a microorganism, usually intentionally.

K/T boundary The Cretaceous–Tertiary boundary (K for Cretaceous, because "C" was already taken for Carboniferous). Boundary where fossils indicate that three-quarters of the animals of that time became extinct, 65 million years ago.

landfill A large hole in the ground into which waste is placed to decay.

light year The distance that light will travel in one year. Equal to 9.56 trillion kilometers.

magma The naturally occurring, mobile mass of molten rock material generated within or beneath the Earth's surface, and from which igneous rocks are derived.

mantle The region of the Earth's subsurface just above the core.

methane A flammable, explosive, colorless, odorless, tasteless gas. It is formed in marshes and swamps from decaying organic matter, and it is a major explosion hazard in mines.

methanogen A microbe that produces methane, usually from the raw materials carbon dioxide and hydrogen.

meteorite A solid mass of mineral or rock matter that has fallen to the Earth's surface from outer space without being completely vaporized in the atmosphere (could be an asteroid or comet).

microbe Broadly, any microorganism, a single-celled organism. Can include photosynthetic algae and many groups that do not fit within the definition "bacteria".

microbial mat A layered assemblage of microbial communities giving the appearance of a mat. Usually held together with a mineral matrix.

mixotrophy The ability to both use sunlight to gather energy (photosynthesis) and to eat other microbes as a predator.

muon An unstable tiny particle smaller than an atom that lasts about a millionth of a second.

Near-Earth Object (NEO) An object whose orbit brings it across the path of the Earth's orbit.

neutron A subatomic particle found in the nucleus of an atom. It has no charge.

nitrogen fixation The fixation of nitrogen gas from the atmosphere into useful, biologically available nitrogen compounds. Carried out by some groups of microbes.

Oort cloud The cloud of comets presumed to exist in the outer edges of the Solar System.

oxidant A compound capable of oxidizing some compounds it comes in contact with. Often damaging to organic molecules from which life is made.

ozone A molecule made up of three oxygen atoms that absorbs ultraviolet radiation of certain wavelengths. Formed by the action of short wavelength ultraviolet radiation on oxygen, which is made up of two oxygen atoms.

ozone hole The depletion of ozone caused by the build-up of natural or man-made chemicals that can break ozone down to oxygen.

panspermia The idea that life can be transferred between planets, either by solar radiation or on rocks ejected during asteroid or comet impacts.

parsec Some astronomers use the term "parsec". One parsec (or parallax arc second) is the distance that an object has to be at so that the average radius of the Earth's orbit takes up one sixtieth of a degree (or one arc second). A parsec is 3.26 light years.

perigalacticon The place in the galactic journey closest to the center of the Galaxy.

Phanerozoic The time of Earth history starting 600 million years ago and running through to the present day. This eon is subdivided into smaller segments.

photon A packet of energy from which light is made (light has the properties of both particles and waves).

photosynthesis The use of sunlight as a source of energy in living organisms.

plasmid A small circular piece of DNA found in many microbes that has encoded on it pathways for particular metabolic processes. Can sometimes be transferred to other microbes.

protogalaxy The gas cloud from which a galaxy is formed.

protoplanetary disk The disk of gas and coalescing material from which the first planets were formed.

psychrophile Organism able to grow at 0 °C or less and up to about 20 °C with an optimum of less than 15 °C.

rad A rad is a measure of radiation. One rad ("radiation absorbed dose") is equivalent to 100 ergs of energy absorbed per gram of tissue.

red giant A large reddish or orange star in the late stages of its evolution.

RNA The intermediate step between DNA, the genetic storage system, and proteins, which are produced by reading the RNA code. 16s RNA is a particularly well conserved piece of RNA that is used to determine how far apart organisms are on the evolutionary tree by looking at the difference in the 16s RNA between them.

Solar System The system of planets and moons that revolves around our Sun (or systems of other planets that revolve around other stars).

spectrometry Covers a variety of techniques that are used to examine the composition of a substance, either its chemical make-up or its physical properties.

spore A resting state of some microbes that is radiation and desiccation resistant. Allows them to remain dormant for long periods of time in adverse conditions.

stratosphere That part of the atmosphere that extends from about 7 to 30 km altitude. It is the region in which most jet airliners fly.

supernova An exploding star, visible for weeks or months.

thermophile Species that thrive in environments where the temperature is high, typically up to 60 °C.

Traps Large regions of lava flows in Siberia and India that bear testament to vast eruptions occurring over millions of years. The Traps, a Nordic word for "staircase" refers to the staircase-like landscapes that form when these Trap provinces are eroded.

tsunami Giant wave created by an object colliding with the oceans or earthquakes under the sea.

ultraviolet radiation Energetic radiation with a wavelength between 1 and 400 nanometers. Causes damage to DNA, but much of it can be screened by atmospheric ozone.

uniformitarianism The viewpoint that the geological or biological history of Earth has been constant over time.

volcanic explosivity index A scale for measuring volcanic eruptions. The VEI index goes from 0 (mildest) to 8 (worst).

Vostok (lake and Antarctic station) A Russian station on the Antarctic continent at 78° S. The location of a lake at a depth of 3.5 kilometers trapped below the ice.

wavelengths Light has a wavelength like waves on the oceans. The value of the wavelength is the distance between two peaks or two troughs of adjacent waves.

white dwarf A high-density star formed when a low-mass star burns up all of its nuclear fuel and then burns off its outer layers.

Index